Probability and Statistics

for Cambridge International A Level

J. Chambers • J. Crawshaw • P. Balaam

OXFORD

UNIVERSITY PRESS

OXFORD
UNIVERSITY PRESS

Great Clarendon Street, Oxford, OX2 6DP, United Kingdom

Oxford University Press is a department of the University of Oxford.
It furthers the University's objective of excellence in research, scholarship,
and education by publishing worldwide. Oxford is a registered trade mark of
Oxford University Press in the UK and in certain other countries

First published by Nelson Thornes Ltd in 2012
This edition published by Oxford University Press in 2015

British Library Cataloguing in Publication Data
Data available

978-1-4085-1563-1

4

Printed by Multivista Global Ltd

Acknowledgements

Page make-up and illustrations: Tech-Set Ltd, Gateshead

Cover photograph: tolgakolcak/iStockphoto

Although we have made every effort to trace and contact all
copyright holders before publication this has not been possible in all
cases. If notified, the publisher will rectify any errors or omissions at
the earliest opportunity.

Links to third party websites are provided by Oxford in good faith
and for information only. Oxford disclaims any responsibility for
the materials contained in any third party website referenced in
this work.

S2 Contents

Introduction

The *Mathematics for Cambridge International A Level* series has been written specifically for students of Cambridge's 9709 syllabus by an experienced author team in collaboration with examiners who are very familiar with the syllabus and examinations. This means that, no matter which combination of modules you have chosen, the content of this series matches the content of the syllabus exactly and will give you firm guidelines on which to base your studies.

In this book, the content of the Statistics 2 module is divided into six chapters that give a sensible order for your studies. Each chapter begins with a list of objectives showing you exactly what is covered.

The following features help you to understand the concepts of the S2 module and to succeed in your exams.

- The introductions to concepts are accompanied by examples of questions, many of which are actual Cambridge questions, together with their solutions. These show each step of working together with commentary on the reasoning processes involved.

- There are numerous exercises for you to practise what you have learned and develop your skills.

- At the end of each chapter there is a Mixed Exercise with more detailed questions covering the content of the chapter. These questions are similar to those found in exam papers and many are from real exam papers.

- Summaries of key information and formulae are given at the end of each chapter to help you revise what you have covered in the chapter.

- Answers to all questions are provided at the back of the book for you to check your answers to exercises.

- Two sample exam papers have been created in the style of Cambridge's International A Level S2 exams to give you some experience of working a full exam paper.

1 The Poisson distribution

In this chapter you will learn about

- the conditions for a Poisson distribution
- calculating Poisson probabilities
- the mean and variance of the Poisson distribution
- the Poisson distribution as an approximation to the binomial distribution
- the normal distribution, with a continuity correction, as an approximation to the Poisson distribution

THE POISSON DISTRIBUTION

Suppose you know that, on average, 3.2 cars per minute arrive at a petrol station. Is it possible to find the probability that, during a particular minute, exactly 4 cars will arrive?

If certain conditions are satisfied, this kind of situation can be modelled by a **Poisson distribution**. This is a special discrete distribution, named after Simeon Poisson, a French mathematician.

Conditions for a Poisson distribution

Consider the discrete random variable X representing the number of occurrences of a certain event in a given interval of time or space.

X is said to follow a **Poisson distribution** if the events occur:

- independently
- singly
- randomly
- at a constant average (mean) rate

These are some examples of variables which could be modelled by a Poisson distribution:

- the number of emergency calls received by an ambulance control centre in 1 hour
- the number of bacteria in 10 ml of pond water
- the number of planes landing at an airfield in 30 minutes
- the number of cars arriving at a petrol station in 10 minutes
- the number of flaws in $20\,\text{m}^2$ of material
- the number of goals scored in a football match.

If the constant average rate is λ, then the distribution can be written

$$X \sim \text{Po}(\lambda)$$

λ is the Greek letter lambda.

The value of λ is all that is needed to define the Poisson distribution completely. It is the only **parameter** of the distribution.

This denotes that X has a **Poisson distribution** with **mean** λ.

When $X \sim \text{Po}(\lambda)$, the probability of x occurrences of a certain event in a given interval of space or time is given by

$$P(X = x) = e^{-\lambda}\frac{\lambda^x}{x!}, \qquad x = 0, 1, 2, 3, \dots \text{ to infinity}$$

$P(X = x)$ is the probability distribution function of X.

Calculating Poisson probabilities

Calculator note

The formula for calculating $P(X = x)$ involves the **exponential function** e^x.

The exponential function e^x is studied in P3.

Locate the button for e^x on your calculator. It is usually the second function on the $\boxed{\ln}$ button. The value of the constant e can be found by calculating e^1. Try it on your calculator: you should get 2.71828...

In this topic you will need to calculate terms with a negative index, such as:

$e^{-0.6} = 0.5488...$

$e^{-1} = 0.36787...$ Try these on **your** calculator to check that you are using the correct buttons.

$e^{-3.2} = 0.0407...$

Using the formula

$$P(X = x) = e^{-\lambda} \frac{\lambda^x}{x!}$$

so $P(X = 0) = e^{-\lambda} \dfrac{\lambda^0}{0!} = e^{-\lambda}$ $[\lambda^0 = 1 \text{ and } 0! = 1]$

$P(X = 1) = e^{-\lambda} \dfrac{\lambda^1}{1!} = e^{-\lambda} \lambda$ $[\lambda^1 = 1 \text{ and } 1! = 1]$

$P(X = 2) = e^{-\lambda} \dfrac{\lambda^2}{2!}$

$P(X = 3) = e^{-\lambda} \dfrac{\lambda^3}{3!}$

and so on for $x = 4, 5, 6, ...$ There is no upper limit for x.

The **probability distribution** can be written

x	0	1	2	3	...
$P(X = x)$	$e^{-\lambda}$	$e^{-\lambda} \lambda$	$e^{-\lambda} \dfrac{\lambda^2}{2!}$	$e^{-\lambda} \dfrac{\lambda^3}{3!}$	and so on

Returning to the situation at the beginning of the chapter, if you assume that the cars arrive at the petrol station independently, singly and at random, at a constant average rate of 3.2 per minute, then X, the number of cars arriving in 1 minute, has a Poisson distribution with mean 3.2.

This is written $X \sim \text{Po}(3.2)$.

So, the probability that, during a particular minute, exactly 4 cars will arrive is given by

$$P(X = 4) = e^{-3.2} \times \frac{3.2^4}{4!} = 0.1780... = 0.178 \text{ (3 s.f.)}$$

Note that the conditions for a Poisson distribution would not be satisfied if, for example, there were traffic lights nearby that caused vehicles to arrive in groups. The traffic must be smooth flowing, with no queues.

Example 1.1

If $X \sim \text{Po}(4.3)$, find:

(i) $P(X = 6)$ (ii) $P(X = 0)$ (iii) $P(X < 3)$.

$P(X = x) = e^{-\lambda} \dfrac{\lambda^x}{x!}$, with $\lambda = 4.3$, so $P(X = x) = e^{-4.3} \dfrac{4.3^x}{x!}$

Substitute $\lambda = 4.3$ into the formula.

(i) $P(X = 6) = e^{-4.3} \times \dfrac{4.3^6}{6!}$

Substitute $x = 6$.

$\qquad = 0.1191...$

$\qquad = 0.119$ (3 s.f.)

(ii) $P(X = 0) = e^{-4.3} \times \dfrac{4.3^0}{0!}$

Remember that $P(X = 0) = e^{-\lambda}$.

$\qquad = e^{-4.3}$

$\qquad = 0.01356...$

$\qquad = 0.0136$ (3 s.f.)

(iii) $P(X < 3) = P(X = 0) + P(X = 1) + P(X = 2)$

$\qquad = e^{-4.3} + e^{-4.3} \times 4.3 + e^{-4.3} \times \dfrac{4.3^2}{2!}$

Remember that $P(X = 1) = e^{-\lambda} \lambda$.

It is more efficient to take out a factor of $e^{-4.3}$, rather than calculating the individual terms.

$\qquad = e^{-4.3} \left(1 + 4.3 + \dfrac{4.3^2}{2!}\right)$

$\qquad = e^{-4.3} \times 14.545$

$\qquad = 0.1973...$

$\qquad = 0.197$ (3 s.f.)

Example 1.2

The number of amoebas in the water from a particular pond follows a Poisson distribution with a mean of 4 amoebas per 10 ml of pond water. Find the probability that in 10 ml of pond water there are:

(i) exactly 5 amoebas   (ii) at most 2 amoebas   (iii) at least 1 amoeba.

Let X be the number of amoebas in 10 ml of pond water.

$X \sim Po(4)$ and $P(X = x) = e^{-\lambda} \dfrac{\lambda^x}{x!}$, with $\lambda = 4$.

(i) $P(X = 5) = e^{-4} \times \dfrac{4^5}{5!}$

$\qquad = 0.01562...$

$\qquad = 0.0156$ (3 s.f.)

(ii) $P(X \leqslant 2)$

$\qquad = P(X = 0) + P(X = 1) + P(X = 2)$

$\qquad = e^{-4} + e^{-4} \times 4 + e^{-4} \times \dfrac{4^2}{2!}$

$\qquad = e^{-4} \left(1 + 4 + \dfrac{4^2}{2!}\right)$

With practice you should be able to go straight to this line of working.

$\qquad = 0.2381...$

$\qquad = 0.238$ (3 s.f.)

(iii)   P(at least 1 amoeba)

$$= P(X = 1) + P(X = 2) + P(X = 3) + \ldots$$

Since there is no upper limit for X, you need to use the fact that the sum of all the probabilities is 1.

So,   P(at least 1 amoeba) = 1 − P(no amoebas)

$$P(X \geqslant 1) = 1 - P(X = 0)$$

$$= 1 - e^{-4}$$

$$= 0.9816\ldots$$

$$= 0.982 \ (3 \text{ s.f.})$$

Example 1.3

The discrete random variable X follows a Poisson distribution with parameter 5. Calculate:

(i) $P(X > 3)$

(ii) $P(5 < X \leqslant 7)$.

You are given that $\lambda = 5$, so $X \sim \text{Po}(5)$

(i) $P(X > 3) = 1 - P(X \leqslant 3)$

$$= 1 - e^{-5}\left(1 + 5 + \frac{5^2}{2!} + \frac{5^3}{3!}\right)$$

$$= 1 - e^{-5} \times 39.3333\ldots$$

First calculate the terms in the bracket, then multiply by e^{-5}, then subtract your answer from 1.

$$= 1 - 0.2650\ldots$$

Do **not** round any numbers in the calculation before you get to the final answer. Then, unless instructed otherwise, round the final answer to 3 significant figures.

$$= 0.7349\ldots = 0.735 \ (3 \text{ s.f.})$$

(ii) $P(5 < X \leqslant 7)$ To find $P(5 < X \leqslant 7)$, include 7 but do not include 5.

$$= P(X = 6) + P(X = 7)$$

$$= e^{-5}\left(\frac{5^6}{6!} + \frac{5^7}{7!}\right)$$

$$= 0.2506\ldots = 0.251 \ (3 \text{ s.f.})$$

Example 1.4

Weak spots occur in the manufacture of a certain cable, independently and randomly, at an average rate of 2.5 weak spots per 100 m length of cable.

(i)   Find the probability that, in a 100 m length of cable, there will be:

     (a)   no weak spots   (b)   more than 2 weak spots.

Paul cuts 8 lengths of cable, each 100 m long.

(ii)   Find the probability that exactly 2 of them have no weak spots.

Let X be the number of weak spots in 100 m of cable, where $X \sim \text{Po}(2.5)$

(i) (a) $P(X = 0) = e^{-2.5}$

$$= 0.08208\ldots$$

$$= 0.0821 \text{ (3 s.f.)}$$

(b) $P(X > 2)$

$$= 1 - (P(X = 0) + P(X = 1) + P(X = 2))$$

$$= 1 - e^{-2.5}\left(1 + 2.5 + \frac{2.5^2}{2!}\right)$$

$$= 1 - 0.5438\ldots$$

$$= 0.4561\ldots$$

$$= 0.456 \text{ (3 s.f.)}$$

(ii) Let Y be the number of lengths in a sample of 8 that have no weak spots.

Recall: Binomial distribution S1 Chapter 5.

Then $Y \sim \text{B}(n, p)$ with $n = 8$, $p = e^{-2.5}$ (from part (i)) and $q = 1 - p = 1 - e^{-2.5}$

So $P(Y = 2) = \binom{8}{2} \times (e^{-2.5})^2 \times (1 - e^{-2.5})^6$

number of ways to choose 2 from 8   probability of 2 successes   probability of 6 failures

$$P(Y = 2) = 0.1128\ldots$$

$$= 0.113 \text{ (3 s.f.)}$$

Note about accuracy:

If you use calculated values of $e^{-2.5}$ and $(1 - e^{-2.5})$ when finding $P(Y = 2)$, do not use rounded answers. As your final answer is required to three significant figures, you must work with at least four significant figures throughout the calculation.

Example 1.5

A shopkeeper sells electric fans. The demand for fans follows a Poisson distribution with mean 3.2 per week.

(i) Find the probability that the demand is exactly 2 fans in any one week.

(ii) The shopkeeper has 4 fans in his shop at the beginning of a week. Find the probability that this will not be enough to satisfy the demand for fans in that week.

(iii) Given instead that he has n fans in his shop at the beginning of a week, find, by trial and error, the least value of n for which the probability of his not being able to satisfy the demand for fans in that week is less than 0.05.

Cambridge Paper 7 Q6 N05

Let X be the number of requests for fans in a week.

(i) $X \sim \text{Po}(3.2)$

$$P(X = 2) = e^{-3.2} \times \frac{3.2^2}{2!} = 0.2087\ldots = 0.209 \text{ (3 s.f.)}$$

(ii) The shopkeeper will not satisfy the demand if there are more than 4 requests in the week.

$$P(X > 4) = 1 - P(X \leqslant 4)$$

$$= 1 - e^{-3.2}\left(1 + 3.2 + \frac{3.2^2}{2!} + \frac{3.2^3}{3!} + \frac{3.2^4}{4!}\right) \qquad \text{Take out a factor of } e^{-3.2}.$$

$$= 1 - 0.780612...$$

$$= 0.2193...$$

P(shopkeeper will not satisfy the demand) = 0.219 (3 s.f.)

(iii) Suppose he has 5 fans, i.e. $n = 5$.

He will not satisfy the demand if there are more than 5 requests.

$$P(X > 5) = 1 - P(X \leqslant 5)$$

$$= 1 - (P(X \leqslant 4) + P(X = 5)) \qquad \text{You know } P(X \leqslant 4) \text{ from part (ii).}$$

$$= 1 - \left(0.780612... + e^{-3.2} \times \frac{3.2^5}{5!}\right)$$

$$= 1 - 0.89459...$$

$$= 0.10540... > 0.05$$

Suppose he has 6 fans, i.e. $n = 6$.

He will not satisfy the demand if there are more than 6 requests.

$$P(X > 6) = 1 - P(X \leqslant 6)$$

$$= 1 - (P(X \leqslant 5) + P(X = 6))$$

$$= 1 - \left(0.89459... + e^{-3.2} \times \frac{3.2^6}{6!}\right) \qquad \text{You know } P(X \leqslant 5) \text{ from the working shown above.}$$

$$= 1 - 0.95538...$$

$$= 0.0446... < 0.05$$

So the least value of n is 6.

Using the Poisson distribution as a model

If you are deciding whether a Poisson model is appropriate, remember that the events should occur independently, singly and at random and that the average rate should be constant. Sometimes situations do not satisfy all these conditions.

For example, a Poisson distribution may be a good model for the number of cars per minute passing a particular checkpoint on a road when the traffic is free flowing. However, cars would not pass randomly if there are traffic jams. You also need to ensure that the average rate per minute is constant during the time interval being considered. For example, it is unlikely that the average rate during the night is the same as during the day.

The Poisson distribution may be a good model for the number of people entering a shop, but it would not be appropriate if friends or family members enter together as the people would not be entering singly and independently.

The Poisson distribution may be appropriate for the *number of accidents* in a particular factory in a year. However this may not be the case for the *number of people injured in accidents* as injuries will not occur independently if several people are injured in a particular accident.

Example 1.6

The number of elephants seen per day during a safari is denoted by the random variable X, with mean 0.8.

(i) (a) State two conditions for X to have a Poisson distribution.

(b) For one of these conditions explain whether you think it likely that it will be satisfied.

(ii) Assuming that the conditions are satisfied, find the probability that Marie sees no elephants during the four days of her safari.

(i) (a) *Any two of the following would be acceptable:*

If one elephant is seen, this does not imply that another is more likely to be seen than otherwise.

Elephants are seen at random.

The mean number of elephants seen per day is constant.

(b) Elephants may go around in pairs or groups, so if one elephant is seen it might be more likely to see another.

(ii) Let X be the number of elephants Marie sees in a day, where $X \sim Po(0.8)$.
$$P(X = 0) = e^{-0.8}$$
P(she sees no elephants in 4 days) $= (e^{-0.8})^4 = 0.04076\ldots = 0.0408$ (3 s.f.)

Exercise 1a

1 If $X \sim Po(1.2)$, find:

(i) $P(X = 2)$ (ii) $P(X = 5)$

(iii) $P(X \leqslant 2)$ (iv) $P(X > 3)$.

2 If $X \sim Po(3.4)$, find:

(i) $P(X = 0)$ (ii) $P(X \geqslant 1)$

(iii) $P(X < 3)$ (iv) $P(X > 4)$.

3 If $X \sim Po(5)$, find:

(i) $P(1 \leqslant X \leqslant 4)$ (ii) $P(5 < X < 8)$

(iii) $P(3 < X \leqslant 6)$ (iv) $P(1 \leqslant X < 3)$.

4 The discrete random variable Y follows a Poisson distribution with parameter 4. Find:

(i) $P(Y < 3)$ (ii) $P(Y > 3)$

(iii) $P(2 < Y < 5)$.

5 Cars arrive singly, randomly and independently at a large service station at an average rate of 1.8 per minute. Find the probability that in a randomly chosen one minute period the number of cars arriving at the service station is:

(i) 2   (ii) at least 3

(iii) fewer than 5.

6 An insurance company receives an average of 2 claims a week from a particular company. Assuming that the number of claims may be modelled by a Poisson distribution, find the probability that the insurance company receives:

(i) exactly 3 claims in a given week

(ii) more than 4 claims in a given week

(iii) no claims in a given week

(iv) at least one claim each week in 4 consecutive weeks.

7 The number of births announced in the personal column of a local weekly newspaper may be modelled by a Poisson distribution with mean 3.2 per week. Find the probability that, in a particular week, the number of births announced is:

(i) at most 3   (ii) exactly 4

(iii) more than 2.

8   Call-outs at a fire station in a town occur at an average rate of 4.2 per night.

   (i) State what needs to be assumed in order to justify a Poisson model.

   Assume that the number of call-outs follows a Poisson distribution.

   (ii) Find the probability that, on a particular night, there will be:

      (a) 3 or more call-outs

      (b) at most 5 call-outs

      (c) at least 1 call-out.

   (iii) Find the probability that there will be no call-outs on exactly 2 of 6 randomly chosen nights.

9   A car hire firm has 3 cars for hire. The number of cars requested each day follows a Poisson distribution with mean 2.1

   (i) Find the probability that, on any given day:

      (a) exactly 2 cars are requested

      (b) all the cars are in use

      (c) the firm will not be able to meet the demand.

   (ii) Find the probability that at least 1 car is requested each day over a period of 6 days.

10 Along a stretch of motorway, vehicle breakdowns occur at an average rate of 2.4 per day. Stating any necessary assumptions,

   (i) find the probability that there will be more than 4 breakdowns in a given day

   (ii) find, by trial and error, the least integer n such that the probability of more than n breakdowns in a day is less than 0.03

11 It is given that $X \sim \text{Po}(\lambda)$ and $P(X = 0) = 2P(X = 1)$. Find the value of λ.

12 The number of goals scored per match by Fanchester City is represented by a random variable X with mean 2.1.

   (i) State two conditions for X to be modelled by a Poisson distribution.

   Assume now that $X \sim \text{Po}(2.1)$.

   (ii) Find $P(3 < X < 7)$

   (iii) A fan club promises the team a special party if they score at least 1 goal in each of the next 12 matches. Find the probability that the team has the party.

Unit interval

Care must be taken to specify the **unit interval** being considered.

In Example 1.2, the mean number of amoebas in 10 ml of pond water is 4,

   so the number of amoebas in **10 ml** is distributed Po(4).

Now suppose you want to find a probability relating to 5 ml of the pond water.

The mean number of amoebas in 5 ml is 2,   The unit interval is halved, so the mean is halved.

   so the number of amoebas in **5 ml** is distributed Po(2).

Similarly,

   the number of amoebas in **30 ml** is distributed Po(12).

Example 1.7

In a certain school, photocopier breakdowns occur randomly and independently, on average 8 times in a school week from Monday to Friday. Find the probability that there will be:

   (i) 5 breakdowns in a week   (ii) 1 breakdown on Monday

   (iii) 8 breakdowns in a fortnight.

(i) Let X be the number of breakdowns in a **week**, so $X \sim \text{Po}(8)$

Define the variable, stating the unit interval.

$$P(X = 5) = e^{-8} \times \frac{8^5}{5!} = 0.09160\ldots = 0.0916 \text{ (3 s.f.)}$$

(ii) Let D be the number of breakdowns in a **day**.

The mean number of breakdowns in a day is $\frac{8}{5} = 1.6$, so $D \sim \text{Po}(1.6)$

You need to define a new variable. Remember to state how it is distributed.

$$P(D = 1) = e^{-1.6} \times 1.6 = 0.3230\ldots = 0.323 \text{ (3 s.f.)}$$

(iii) Let F be the number of breakdowns in **a fortnight**.

Define another new variable.

The mean number of breakdowns in a fortnight is $8 \times 2 = 16$, so $F \sim \text{Po}(16)$

$$P(F = 8) = e^{-16} \times \frac{16^8}{8!} = 0.01198\ldots = 0.0120 \text{ (3 s.f.)}$$

Example 1.8

At a certain airfield planes land at random times at a constant average rate of one every 10 minutes.

(i) Find the probability that exactly 5 planes will land in a period of one hour.
(ii) Find the probability that at least 2 planes will land in a period of 16 minutes.
(iii) Given that 5 planes landed in an hour, calculate the conditional probability that 1 plane landed in the first half hour and 4 in the second half hour.

Cambridge Paper 7 Q6 J04

The average number of planes in 10 minutes is 1, so the average rate is 0.1 per minute.

(i) Let X be the number of planes landing in **1 hour**.

The average number of planes in 1 hour is $60 \times 0.1 = 6$, so $X \sim \text{Po}(6)$

$$P(X = 5) = e^{-6} \times \frac{6^5}{5!} = 0.16062\ldots = 0.161 \text{ (3 s.f.)}$$

(ii) Let Y be the number of planes landing in **16 minutes**.

The average number of planes in 16 min is $16 \times 0.1 = 1.6$, so $Y \sim \text{Po}(1.6)$

$$
\begin{aligned}
P(Y \geqslant 2) &= 1 - (P(Y = 0) + P(Y = 1)) \\
&= 1 - e^{-1.6}(1 + 1.6) \\
&= 0.47506\ldots \\
&= 0.475 \text{ (3 s.f.)}
\end{aligned}
$$

(iii) Let L be the number of planes landing in **30 minutes**, where $L \sim \text{Po}(30 \times 0.1)$, i.e. $L \sim \text{Po}(3)$

In the first half-hour, $\qquad P(L = 1) = e^{-3} \times 3$

In the second half-hour, $\qquad P(L = 4) = e^{-3} \times \frac{3^4}{4!}$

So, P(1 in first half-hour and 4 in second half-hour)

For independent events
$P(A \text{ and } B) = P(A) \times P(B)$

$$= (e^{-3} \times 3) \times \left(e^{-3} \times \frac{3^4}{4!}\right) = 0.025097\ldots$$

P(1 in first half-hour and 4 in second half-hour, given 5 in an hour)

$$= \frac{P(1 \text{ in first half-hour and 4 in second half-hour})}{P(5 \text{ in an hour})}$$

Conditional probability S1 Chapter 3.

$$= \frac{0.025097\ldots}{0.16062\ldots} \longleftarrow \text{From part (i)}$$

$$= 0.1562\ldots$$

$$= 0.156 \text{ (3 s.f.)}$$

Note: You could get an <u>exact</u> answer to part (iii) by simplifying as follows:

P(1 in first half-hour and 4 in second half-hour),

$$= e^{-3} \times 3 \times e^{-3} \times \frac{3^4}{4!} = e^{-6} \times \frac{3^5}{4!}$$

P(1 in first half-hour and 4 in second half-hour, given 5 in 1 hour)

$$= \frac{e^{-6} \times \dfrac{3^5}{4!}}{e^{-6} \times \dfrac{6^5}{5!}} = \left(\frac{1}{2}\right)^5 \times 5 = \frac{5}{32}$$

$$\frac{3^5}{6^5} = \left(\frac{3}{6}\right)^5 = \left(\frac{1}{2}\right)^5$$

$$\frac{1}{4!} \div \frac{1}{5!} = \frac{5!}{4!} = 5$$

The Poisson probability distribution

These vertical line diagrams show the probability distribution of X, where $X \sim Po(\lambda)$, for various values of λ. Notice that the distribution has a tail to the right (a positive skew) for small values of λ but becomes closer to being symmetrical as the value of λ increases.

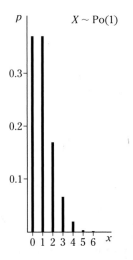

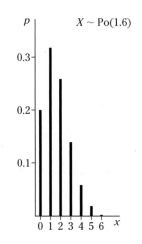

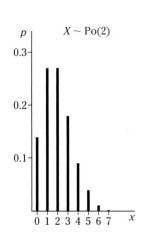

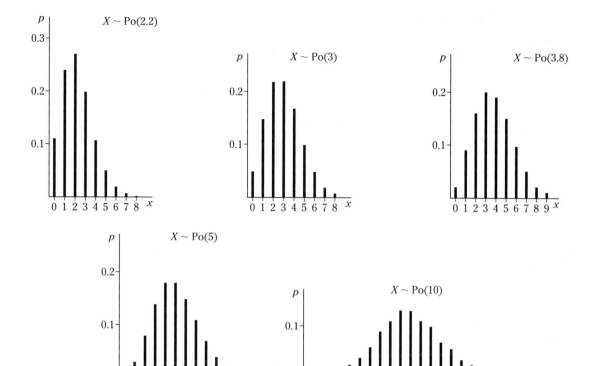

The mode of the Poisson distribution

The mode is the value of X that is most likely to occur, i.e. the one with the greatest probability.

From the diagrams, you can see that:

- when $\lambda = 1$, there are two modes, 0 and 1
- when $\lambda = 2$, there are two modes, 1 and 2
- when $\lambda = 3$, there are two modes, 2 and 3.

In general, when λ is a whole number, there are two modes, $\lambda - 1$ and λ.

For example, if $X \sim \text{Po}(8)$, the modes are 7 and 8.

Notice also that:

- when $\lambda = 1.6$, the mode is 1
- when $\lambda = 2.2$, the mode is 2
- when $\lambda = 3.8$, the mode is 3.

In general, when λ is not a whole number, the mode is the next whole number below λ.

For example, if $X \sim \text{Po}(4.9)$, the mode is 4.

Note: Knowledge of the mode will not be needed in the examination.

The mean and variance of the Poisson distribution

Since λ is the average rate of occurrences, the mean, $E(X)$, is λ.
In a Poisson distribution, the variance is the same as the mean, so $\text{Var}(X)$ is also λ.

Recall: $E(X)$ and $\text{Var}(X)$ for discrete probability distributions (S1 Chapter 4).

If X follows a **Poisson** distribution such that $X \sim \text{Po}(\lambda)$, then:

mean $\mu = \text{E}(X) = \lambda$

variance $\sigma^2 = \text{Var}(X) = \lambda$

standard deviation $\sigma = \sqrt{\lambda}$ Recall: standard deviation $= \sqrt{\text{variance}}$

Notation

In the **List of Formulae** provided in the examination, a is used instead of λ and the theory relating to the Poisson distribution is summarised as follows:

For the Poisson distribution $\text{Po}(a)$:

$$p_r = \text{e}^{-a} \frac{a^r}{r!} \qquad\qquad \mu = a \qquad\qquad \sigma^2 = a$$

Example 1.9

The number of accidents per day at a large factory follows a Poisson distribution with parameter 2.25.

(i) State the mean and standard deviation of the number of accidents per day.

(ii) On how many days in November would you expect the number of accidents to be within one standard deviation of the mean?

Let X be the number of accidents in a day, where $X \sim \text{Po}(\lambda)$

(i) Mean $= \lambda = 2.25$

Standard deviation $= \sqrt{\lambda} = \sqrt{2.25} = 1.5$

(ii) $X \sim \text{Po}(2.25)$

First find the probability that X is within one standard deviation of the mean.

$\mu - \sigma = 2.25 - 1.5 = 0.75$

$\mu + \sigma = 2.25 + 1.5 = 3.75$

$\text{P}(\mu - \sigma < X < \mu + \sigma)$

$\qquad = \text{P}(0.75 < X < 3.75)$

$\qquad = \text{P}(X = 1) + \text{P}(X = 2) + \text{P}(X = 3)$

$\qquad = \text{e}^{-2.25} \left(2.25 + \frac{2.25^2}{2!} + \frac{2.25^3}{3!} \right)$

$\qquad = 0.7040\ldots$

Now multiply this probability by the number of days in November.

$\qquad 0.07040\ldots \times 30 = 21.12\ldots$

So, you would expect the number of accidents to be within one standard deviation of the mean on approximately 21 days in November.

Using $\log_e x$ or $\ln x$

Questions on the Poisson distribution may involve the use of **logarithms to the base e** (ln), studied in P3.

To find $\log_e$ of a number use $\boxed{\ln}$ on your calculator. You should find, for example, that $\log_e(5.4) = \ln(5.4) = 1.6863\ldots$

The function $\log_e x$ or $\ln x$ is the **inverse** function of e^x.

To find x when you are given e^x, take $\log_e$ of both sides, for example:

If $\qquad e^x = 0.315$

then $\qquad \ln(e^x) = \ln(0.315)$ $\qquad\qquad$ Note: $\ln(e^x) = x$

$\qquad\qquad x = -1.15518\ldots$

To check, if you use the number on the screen you will find that $e^{-1.15518\ldots} = 0.315$

It is important that you can use the $\boxed{\ln}$ function, so practise with these:

$\qquad$ Find x and then do a check, as above, for $e^x = 2.9$, $e^x = 0.05$, $e^x = 17.8$

Example 1.10

The random variable X is distributed Po($0.3n$).

(i) Find $P(X \geqslant 1)$ when $n = 4$.

(ii) Find the least integer value of n so that $P(X \geqslant 1)$ is greater than 0.95

$X \sim \text{Po}(0.3n)$

(i) $P(X \geqslant 1) = 1 - P(X = 0)$

$\qquad\qquad\quad = 1 - e^{-0.3n}$

When $n = 4$,

$\quad P(X \geqslant 1) = 1 - e^{-0.3 \times 4}$

$\qquad\qquad\quad = 1 - e^{-1.2}$

$\qquad\qquad\quad = 0.6988\ldots$

$\qquad\qquad\quad = 0.699 \text{ (3 s.f.)}$

(ii) You want $P(X \geqslant 1) > 0.95$

$\quad$ i.e. $\quad 1 - e^{-0.3n} > 0.95$

$\quad$ Rearrange:

$\qquad\qquad 1 - 0.95 > e^{-0.3n}$

$\qquad\qquad\quad 0.05 > e^{-0.3n}$

so $\qquad\quad e^{-0.3n} < 0.05$

$\quad$ Take $\log_e$ of both sides:

$\qquad\qquad -0.3n < \ln(0.05)$

$\qquad\qquad -0.3n < -2.9957\ldots$

You could use trial and improvement to find n, but this may be more time-consuming than using logs.

Now divide by -0.3

Remember that dividing by a negative quantity reverses the inequality:

$$n > \frac{-2.9957\ldots}{-0.3}$$

$$n > 9.985\ldots$$

So the least integer value of n is 10.

Check:

When $n = 9$, $P(X \geqslant 1) = 1 - e^{-2.7} = 0.9327\ldots < 0.95$

When $n = 10$, $P(X \geqslant 1) = 1 - e^{-3} = 0.9502\ldots > 0.95$

Example 1.11

The number of emergency telephone calls to the electricity board office in a certain area in t minutes is known to follow a Poisson distribution with mean $\frac{1}{80}t$.

(i) Find the probability that there will be at least 3 emergency telephone calls to the office in any 20-minute period.

(ii) The probability that no emergency telephone call is made to the office in a period of k minutes is 0.9. Find k.

Cambridge Paper 7 Q4 N03

(i) Let X be the number of telephone calls in 20 minutes.

Then $X \sim \text{Po}\left(\frac{1}{80} \times 20\right)$, i.e. $X \sim \text{Po}(0.25)$

$$P(X \geqslant 3) = 1 - P(X \leqslant 2)$$

$$= 1 - e^{-0.25}\left(1 + 0.25 + \frac{0.25^2}{2!}\right)$$

$$= 1 - 0.9978\ldots$$

$$= 0.002161\ldots$$

$$= 0.00216 \text{ (3 s.f.)}$$

(ii) Let Y be the number of telephone calls in k minutes.

Then $Y \sim \text{Po}\left(\frac{1}{80} \times k\right)$, i.e. $Y \sim \text{Po}\left(\frac{k}{80}\right)$

$$P(Y = 0) = e^{-\frac{k}{80}}$$

But you are given that $P(Y = 0) = 0.9$

Therefore $\quad e^{-\frac{k}{80}} = 0.9$

Take $\log_e$ of both sides:

$$-\frac{k}{80} = \ln(0.9)$$

$$= -0.1053\ldots$$

$$k = 80 \times 0.1053\ldots$$

$$= 8.4288\ldots$$

$$= 8.43 \text{ (3 s.f.)}$$

Exercise 1b

1 Emails arrive randomly on Oliver's computer at an average rate of 1.46 per hour. Stating a necessary assumption, find the probability that between 9.00am and 11.30am Oliver receives:

  (i) no emails   (ii) at most 4 emails.

2 During each afternoon, helicopters land on a holiday island one at a time at a constant average rate of 1 every 20 minutes. Find the probability that, on a particular afternoon:

  (i) exactly 2 helicopters land between 2.30pm and 3.30pm

  (ii) at least 2 helicopters land between 3.40pm and 4.20pm

  (iii) no helicopters land between 2.00pm and 2.15pm.

3 The number of telephone calls made to a large office between 9.30am and 10.30am on a weekday follows a Poisson distribution with mean 6.

  Find the probability that:

  (i) there will be 2 or more calls between 9.30am and 10.30am on Tuesday

  (ii) there will be exactly 2 calls between 9.30am and 9.40am on Wednesday

  (iii) during a period of 5 days there will be exactly 3 days when there are no calls between 10.00am and 10.10am.

4 The number of letters advertising various products that Alex receives per day may be modelled by a Poisson distribution with parameter 0.5

  Find the probability that Alex receives:

  (i) exactly 2 letters on a particular day

  (ii) no letters on 3 consecutive days

  (iii) more than 4 letters in a 6-day period.

5 The number of bacterial colonies on a Petri dish can be modelled by a Poisson distribution with average number 2.5 per cm^2. Find the probability that:

  (i) there are no bacterial colonies in $1\,cm^2$

  (ii) there are more than 4 bacterial colonies in $2\,cm^2$

  (iii) there are exactly 6 bacterial colonies in $4\,cm^2$.

6 In a factory a particular machine breaks down at an average rate of 3.2 times a week. The number of breakdowns may be modelled by a Poisson distribution.

  (i) Find the probability that the number of breakdowns is:

    (a) at most 2 in a given week

    (b) more than 5 in a given fortnight.

  (ii) Given that the machine broke down 7 times in a fortnight, calculate the conditional probability that the machine broke down 3 times in the first week and 4 times in the second week.

7 In the manufacture of commercial carpet, small faults occur in the carpet independently and at random at an average rate of 0.95 per 20 square metres.

  Find the probability that in a randomly selected 20 square metres of this carpet there are:

  (i) no faults   (ii) at most 2 faults.

  A new office block has 10 rooms. Each room has a floor area of 80 square metres and is carpeted using the commercial carpet described above.

  (iii) For any one of these rooms, find the probability that the carpet in the room contains:

    (a) at least 2 faults

    (b) exactly 3 faults

    (c) at most 5 faults.

  (iv) Find the probability that in exactly 5 of the rooms in the office block the carpets will contain exactly 3 faults.

8 The random variable X follows a Poisson distribution with parameter 6.3

  (i) Find   (a) $P(X = 5)$ (b) $P(X = 6)$
        (c) $P(X = 7)$.

  (ii) What is the most likely value of X?

9 $X \sim Po(5.3)$

  (i) State the variance of X.

  (ii) Find $P(X < \sigma)$ where σ is the standard deviation of X.

10 A shop sells printers. The number of printers sold per week may be modelled by a Poisson distribution with standard deviation 2.3

  Find:

  (i) the mean number of printers sold in a week

    (ii) the probability that fewer than the mean number of printers are sold in a week

    (iii) exactly 8 printers are sold in a fortnight.

11 The number of telephone calls made to a health centre may be modelled by a Poisson distribution. The probability that there are exactly 2 calls during a 5-minute interval is the same as the probability that there are exactly 3 calls during a 5-minute interval.

    (i) Find the mean number of calls in a 5-minute interval.

    (ii) Find the standard deviation of the number of calls in a 10-minute interval.

12 The random variable X follows a Poisson distribution. Given that $\text{Var}(X) = 4.2$, find:

    (i) $P(X < E(X))$

    (ii) the standard deviation of X.

13 The random variable X is distributed $\text{Po}(1.4t)$.

    (i) If $t = 2$, find $P(X = 4)$.

    (ii) If $t = 4$, find $P(X < 2)$.

14 The random variable X is distributed $\text{Po}(0.5t)$.

Giving your answers to 3 significant figures, find

    (i) the value of t when $P(X = 0) = 0.04$

    (ii) the value of t when $P(X \geqslant 1) = 0.9$

15 It is given that $X \sim \text{Po}(0.8n)$, where n is a positive whole number. Find the least value of n such that $P(X = 0) < 0.1$

THE POISSON DISTRIBUTION AS AN APPROXIMATION TO THE BINOMIAL DISTRIBUTION

In certain circumstances, binomial probabilities can be calculated approximately by using the Poisson distribution.

In this section you need to be familiar with the theory of the binomial distribution covered in S1, especially these important results:

If X follows a **binomial** distribution $B(n, p)$,

then $E(X) = \mu = np$

        $\text{Var}(X) = \sigma^2 = npq$ where $q = 1 - p$

        Standard deviation $\sigma = \sqrt{npq}$

The Poisson approximation

When n is large ($n > 50$) and p is small enough to ensure that $np < 5$, the binomial distribution can be approximated by a **Poisson distribution** with the *same mean*, i.e. a Poisson distribution with mean np.

When $n > 50$ and $np < 5$,

    $X \sim B(n, p)$

can be approximated by

    $X \sim \text{Po}(np)$ $\lambda = np$

The larger the value of n and the smaller the value of p, the better the approximation.

Example 1.12

In a certain city the probability that a person has blood type AB is 0.035

A random sample of 120 people is selected from the city. By using a suitable approximation, find the probability that:

    (i) exactly 5 people in the group have blood type AB

    (ii) at least 2 people in the group have blood type AB.

Let X be the number of people in 120 with blood type AB.

$X \sim \text{B}(120, 0.035)$ with $n = 120$ and $p = 0.035$

$np = 120 \times 0.035 = 4.2$

> Always define the distribution as binomial before using the Poisson approximation.

Since $n > 50$ and $np < 5$, use the Poisson approximation, where $X \sim \text{Po}(4.2)$

Using the binomial distribution directly:

(i) $\text{P}(X = 5) = e^{-4.2} \times \dfrac{4.2^5}{5!}$

$= 0.1633...$

$= 0.163$ (3 s.f.)

(i) $\text{P}(X = 5) = \dbinom{120}{5} \times 0.035^5 \times 0.965^{115} = 0.166$ (3 s.f.)

(ii) $\text{P}(X \geqslant 2) = 1 - \text{P}(X \leqslant 1)$

$= 1 - e^{-4.2}(1 + 4.2)$

$= 0.9220...$

$= 0.922$ (3 s.f.)

(ii) $\text{P}(X \geqslant 2) = 1 - \text{P}(X \leqslant 1)$

$= 1 - (0.965^{120} + 120 \times 0.035 \times 0.965^{119})$

$= 1 - 0.0744...$

$= 0.9255...$

$= 0.926$ (3 s.f.)

The answers obtained using the Poisson approximation are comparable with those obtained directly using the binomial distribution and the calculations are much simpler.

Example 1.13

1.5% of the population of the UK can be classified as 'very tall'.

(i) The random variable X denotes the number of people in a sample of n people who are classified as very tall. Given that $\text{E}(X) = 2.55$, find n.

(ii) By using the Poisson distribution as an approximation to a binomial distribution, calculate an approximate value for the probability that a sample of size 210 will contain fewer than 3 people who are classified as very tall.

Cambridge Paper 7 Q2 N02

(i) $X \sim \text{B}(n, p)$, where $p = 1.5\% = 0.015$

$\text{E}(X) = np = n \times 0.015 = 0.015n$

so $0.015n = 2.55$

$n = \dfrac{2.55}{0.015} = 170$

(ii) Let X be the number of people classified as very tall in a sample of 210 people.

$X \sim \text{B}(210, 0.015)$ with $n = 210$ and $p = 0.015$

$np = 210 \times 0.015 = 3.15$

> Remember to justify the use of the Poisson approximation.

Since $n > 50$ and $np < 5$, use the Poisson approximation, where $X \sim \text{Po}(3.15)$

$\text{P}(X < 3) = \text{P}(X = 0) + \text{P}(X = 1) + \text{P}(X = 2)$

$= e^{-3.15}\left(1 + 3.15 + \dfrac{3.15^2}{2!}\right)$

$= 0.3904...$

$= 0.390$ (3 s.f.)

Note: If you are asked to use a Poisson approximation but instead use the binomial distribution, you will not be awarded full marks, even though your answer will be more accurate.

Example 1.14

The probability that a particular type of cell phone works properly when it is first used is 0.97

During a busy weekend, a store sells 150 of this type of cell phone. Using a suitable approximation, find the probability that more than 146 of the cell phones work properly when they are first used.

Let X be the number of cell phones in 150 that work properly when they are first used.

$X \sim B(150, 0.97)$

$n > 50, np = 150 \times 0.97 = 145.5$

Since np is not less than 5, the conditions for the Poisson approximation are not satisfied.

Consider instead the variable Y, where Y is the number of cell phones that do **not** work properly.

This is a useful method when n is large and p is close to 1.

Then $Y \sim B(150, 0.03)$ with $n = 150$ and $p = 0.03$

$np = 150 \times 0.03 = 4.5$

Since $n > 50$ and $np < 5$, use $Y \sim Po(4.5)$

$P(X > 146)$

$= P(X = 147) + P(X = 148) + P(X = 149) + P(X = 150)$

$= P(Y = 3) + P(Y = 2) + P(Y = 1) + P(Y = 0)$

$= P(Y \leq 3)$

$= e^{-4.5} \left(1 + 4.5 + \frac{4.5^2}{2!} + \frac{4.5^3}{3!}\right)$

$= 0.3422...$

$= 0.342$ (3 s.f.)

Exercise 1c

Use a suitable approximation where appropriate.

1 The random variable X follows a binomial distribution, where $X \sim B(100, 0.03)$

Find the following probabilities using:

(i) the binomial distribution

(ii) a suitable Poisson approximation.

(a) $P(X = 0)$ (b) $P(X = 2)$

(c) $P(X = 4)$.

2 Calculate the following probabilities, using a suitable approximation where appropriate.

(i) $P(X > 3)$ when $X \sim B(350, 0.004)$

(ii) $P(X \leq 2)$ when $X \sim B(40, 0.15)$.

3 The probability that a bolt is defective is 0.2%

Bolts are packed in boxes of 500.

(i) Find the probability that in a randomly chosen box:

(a) there are 2 defective bolts

(b) there are more than 3 defective bolts.

(ii) Two boxes are picked at random from the production line. Find the probability that one has 2 defective bolts and the other has no defective bolts.

(iii) Three boxes are selected at random from the production line. Find the probability that they contain no defective bolts.

4 On average 1 in 200 cars develops a particular fault within a year of purchase.

(i) Find the probability that none of a randomly chosen sample of 250 cars develops the fault within a year of purchase.

(ii) Find the probability that, in a randomly chosen sample of 300 cars, exactly 2 develop the fault within a year of purchase.

5   Two fair cubical dice are thrown.

(i) When both dice show 6, it is called a 'double six'. What is the probability of throwing a double six?

Two fair cubical dice are thrown a total of 90 times.

(ii) Find the probability that at least 2 double sixes are thrown.

6   A newspaper reports that 8.6% of people in a particular region are left-handed. A random sample of 55 people in the region is selected. Use a suitable approximation to find the probability that fewer than 4 people in the sample are left-handed.

7   It is known that the probability that a person has blood type X is $\frac{1}{80}$. Using a suitable approximation, find the probability that there are at least 5 people with blood type X in a random sample of 200 people.

8   The probability that I dial a wrong number when making a telephone call is 0.018

In a typical week I make 50 telephone calls.

(i) Using a Poisson approximation to a binomial model, find the probability that in a given week:

(a) I dial no wrong numbers

(b) I dial more than 2 wrong numbers.

(ii) Comment on the suitability of the binomial model and of the Poisson approximation.

9   A manufacturer produces jelly beans in a variety of colours. On average 2.5% of the jelly beans are purple. The jelly beans are randomly placed into cartons, each containing 150 jelly beans.

(i) A carton is selected at random. Find the probability that the carton contains:

(a) no purple jelly beans

(b) at least 4 purple jelly beans.

(ii) Two cartons are selected at random. Find the probability that both cartons contain at least 1 purple jelly bean.

10   A nurseryman found that on average 98% of sunflower seeds germinate. Using a suitable approximation, find the probability that in a pack containing 150 sunflower seeds:

(i) more than 4 do not germinate

(ii) at least 145 germinate.

11   An aircraft has 240 seats. Past records show that, on average, 2% of people who have booked tickets for a flight do not turn up. The airline sells 244 tickets for a particular flight. Using a suitable approximation, find the probability that:

(i) exactly 4 people do not turn up for the flight

(ii) there are empty seats on the flight

(iii) there are not enough seats for everyone who turns up for the flight.

12   It is known that a particular allergy affects 1.5% of children under 10 years old. The random variable X is the number of children who suffer from the allergy in a random sample of 300 children under 10 years old.

(i) Write down the probability distribution for X.

(ii) Using a suitable approximation, find:

(a) $P(X = 3)$ (b) $P(X \geqslant 5)$.

(iii) How many children under 10 years old must be sampled in order that the probability that no children suffer from the allergy is less than 5%?

THE NORMAL DISTRIBUTION AS AN APPROXIMATION TO THE POISSON DISTRIBUTION

In this section you need to be familiar with finding normal probabilities and using continuity corrections studied in S1 Chapter 6.

In certain circumstances, a normal approximation can be used when calculating probabilities for a Poisson distribution.

The normal approximation

You can see from the probability diagrams on pages 10–11 that as λ increases the Poisson distribution becomes less skewed and closer to being symmetrical.

This diagram shows the probability distribution of X, where $X \sim \text{Po}(15)$. The distribution is almost symmetrical and bell-shaped.

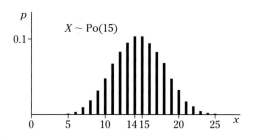

In fact, when λ is large, the distribution takes on the characteristic normal shape and the Poisson distribution can be approximated by a **normal distribution** with the same mean and the same variance as the Poisson distribution.

Now when $X \sim \text{Po}(\lambda)$, the mean and the variance are both equal to λ.

So, when λ is large,

$$X \sim \text{Po}(\lambda)$$

can be approximated by

$$X \sim \text{N}(\lambda, \lambda), \quad \text{with } \mu = \lambda \text{ and } \sigma = \sqrt{\lambda}$$

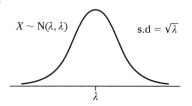

Generally $\lambda > 15$ gives reasonable results, but the larger the value of λ, the better the approximation.

Continuity correction

As with the normal approximation to the binomial distribution studied in S1, a **continuity correction** is required, since you are using a continuous distribution as an approximation to a discrete distribution. This involves adding or subtracting 0.5 as appropriate and is illustrated in the following examples.

Example 1.15

A retailer sells televisions. The number of televisions sold per week follows a Poisson distribution with mean 45. Find the probability that the number of televisions sold in a particular week:

    (i) will exceed 60   (ii) is at least 34 but fewer than 50.

X is the number of televisions sold in a week.

$X \sim \text{Po}(45)$

$\lambda = 45 > 15$, so use a normal approximation, where

$X \sim \text{N}(45, 45)$ with $\mu = 45$ and $\sigma = \sqrt{45}$

> Do not approximate here as this will lead to inaccuracy. Use $\sqrt{45}$ on the calculator.

  (i) $P(X > 60) \rightarrow P(X > 60.5)$

> Apply a continuity correction.

$$= P\left(Z > \frac{60.5 - 45}{\sqrt{45}}\right)$$

$$= P(Z > 2.311)$$

$$= 1 - \Phi(2.311)$$

$$= 1 - 0.9896$$

$$= 0.0104 \text{ (3 s.f.)}$$

S1 Chapter 6
$P(Z > a) = 1 - \Phi(a)$

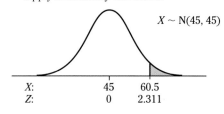

(ii) P(X is at least 34 but fewer than 50)

$$= P(34 \leqslant X < 50)$$

$$\rightarrow P(33.5 < X < 49.5)$$

$$= P\left(\frac{33.5 - 45}{\sqrt{45}} < Z < \frac{49.5 - 45}{\sqrt{45}}\right)$$

$$= P(-1.714 < Z < 0.671)$$

$$= \Phi(0.671) - (1 - \Phi(1.714))$$

$$= 0.7489 - (1 - 0.9568)$$

$$= 0.7057$$

$$= 0.706 \text{ (3 s.f.)}$$

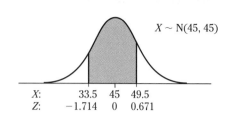

$X:$ 33.5   45   49.5
$Z:$ −1.714   0   0.671

$P(-a < Z < b) = \Phi(b) - (1 - \Phi(a))$

Example 1.16

It is proposed to model the number of people per hour calling a car breakdown service between the times 09 00 and 21 00 by a Poisson distribution.

 (i) Explain why a Poisson distribution may be appropriate for this situation.

People call the car breakdown service at an average rate of 20 per hour, and a Poisson distribution may be assumed to be a suitable model.

 (ii) Find the probability that exactly 8 people call in any half hour.

(iii) By using a suitable approximation, find the probability that exactly 250 people call in the 12 hours between 09 00 and 21 00.

Cambridge Paper 7 Q5 J07

 (i) A Poisson distribution may be appropriate provided that people call randomly, independently and at a constant average rate per hour throughout the time period from 09 00 to 21 00.

(ii) The average rate of calls in an hour is 20, so the average rate in a **half-hour** period is 10.

Let X be the number of calls in a half-hour period, so $X \sim Po(10)$

$$P(X = 8) = e^{-10} \times \frac{10^8}{8!} = 0.1125... = 0.113 \text{ (3 s.f.)}$$

(iii) Let Y be the number of calls between 09 00 and 21 00. The time period is **12 hours**, so $Y \sim Po(240)$

Define the variable as Poisson first. Then check the conditions for a normal approximation.

Since $\lambda > 15$, use the normal approximation, where $Y \sim N(240, 240)$ with $\mu = 240$ and $\sigma = \sqrt{240}$

$$P(Y = 250) \rightarrow P(249.5 < Y < 250.5)$$

Apply continuity corrections.

$$= P\left(\frac{249.5 - 240}{\sqrt{240}} < Z < \frac{250.5 - 240}{\sqrt{240}}\right)$$

$$= P(0.613 < Z < 0.678)$$

$$= \Phi(0.678) - \Phi(0.613)$$

$P(a < Z < b) = \Phi(b) - \Phi(a)$

$$= 0.7512 - 0.7301$$

$$= 0.0211 \text{ (3 s.f.)}$$

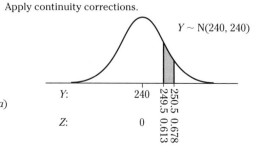

$Y \sim N(240, 240)$

$Y:$ 240   250.5   0.678 / 249.5   0.613
$Z:$ 0

Exercise 1d

1 A random variable X has the distribution Po(64).

(i) State the mean and standard deviation of an appropriate approximating distribution.

Using this approximating distribution, find:

(ii) $P(X \leqslant 52)$ (iii) $P(X > 44)$.

2 If $X \sim$ Po(24), use a suitable approximation to find:

(i) $P(X \leqslant 25)$ (ii) $P(22 \leqslant X \leqslant 26)$

(iii) $P(X > 23)$.

3 If $X \sim$ Po(35), use a suitable approximation to find:

(i) $P(X \leqslant 33)$ (ii) $P(33 < X < 37)$

(iii) $P(X > 37)$ (iv) $P(X = 37)$.

4 If $X \sim$ Po(60), use a suitable approximation to find:

(i) $P(50 < X \leqslant 58)$ (ii) $P(57 \leqslant X < 68)$

(iii) $P(X > 52)$ (iv) $P(X \geqslant 70)$.

5 A radioactive disintegration gives counts that follow a Poisson distribution with a mean count of 25 per second. Use a suitable approximation to find the probability that:

(i) the count is at least 23 but not more than 27 in a one second interval

(ii) the count is greater than 120 in a five second interval.

6 The number of calls per hour received by an office switchboard follows a Poisson distribution with parameter 30. Using the normal approximation to the Poisson distribution, find the probability that, in one hour:

(i) there are more than 33 calls

(ii) there are between 25 and 28 calls (inclusive)

(iii) there are exactly 34 calls.

7 In a certain factory the number of accidents occurring in a month follows a Poisson distribution with standard deviation 2. Use a suitable approximation to find the probability that there will be:

(i) at least 40 accidents during 1 year

(ii) 20 or fewer accidents in 6 months.

8 The number of bacteria on a plate viewed under a microscope follows a Poisson distribution with parameter 60. A plate is chosen at random.

(i) Find the probability that there are at least 70 bacteria on the plate.

(ii) Find the probability that there are more than 54 but fewer than 75 bacteria on the plate.

A plate is rejected if fewer than 38 bacteria are found.

(iii) Find the probability that a randomly chosen plate will be rejected.

(iv) If 2000 such plates are viewed, how many would you expect to be rejected?

9 In an experiment with a radioactive substance, the number of particles reaching a counter over a given period of time follows a Poisson distribution with mean 22.

Find the probability that the number of particles reaching the counter over the given period of time is:

(i) at most 21   (ii) 18 or more

(iii) exactly 22.

10 Accidents on a particular railway line occur at an average rate of 1 every 2 months. Find the probability that there are:

(i) 25 or more accidents in 4 years

(ii) 30 or fewer accidents in 5 years

(iii) exactly 18 accidents in 3 years.

11 The number of eggs laid by a particular type of insect follows a Poisson distribution with parameter 200. Find the probability that the number of eggs laid by an insect of this type is:

(i) more than 170

(ii) fewer than 205

(iii) at least 180 but not more than 240.

12 A grain store containing wheat has been contaminated by weed seeds. The farmer finds that there are, on average, 4 weed seeds per kilogram of wheat. Assuming that the number of weed seeds may be modelled by a Poisson distribution, find the probability that 20 kg of wheat contains:

  (i) fewer than 60 weed seeds

  (ii) at least 70 weed seeds.

13 In spring, the number of tadpoles in a particular pond follows a Poisson distribution with a mean of 2 tadpoles per 100 ml of pond water. Jack collects 1 litre of pond water in his bucket and examines it for tadpoles. Find the probability that the bucket will contain:

  (i) more than 16 tadpoles

  (ii) fewer than 14 tadpoles

  (iii) between 24 and 30 tadpoles (inclusive).

14 The number of faults per square metre of a particular type of cloth is denoted by X.

  (i) State conditions that must be satisfied for X to follow a Poisson distribution.

Now assume that X has the distribution Po(0.8).

  (ii) Find the probability that there are

    (a) no faults in $2\,m^2$ of cloth,

    (b) at least 2 faults in $3\,m^2$ of cloth.

  (iii) Cloth is stored on rolls, each containing $100\,m^2$ of cloth. Find the probability that a roll of cloth contains more than 100 faults.

Summary

Conditions for a Poisson distribution

The random variable X is the number of occurrences of a certain event in a given interval of space or time.

$X \sim$ Po(λ) when events occur:

- independently
- singly
- randomly
- at a constant average (mean) rate λ.

Poisson probability distribution

When $X \sim$ Po(λ), the probability of x occurrences in a given interval is given by

$$P(X = x) = e^{-\lambda}\frac{\lambda^x}{x!}, \qquad x = 0, 1, 2, 3, \ldots \text{ to infinity}$$

Mean and variance of the Poisson distribution

    Mean $\mu = E(X) = \lambda$

    Variance $\sigma^2 = \text{Var}(X) = \lambda$

    Standard deviation $\sigma = \sqrt{\lambda}$

Poisson approximation to the binomial distribution

When $n > 50$ and $np < 5$,

    $X \sim$ B(n, p)

can be approximated by

    $X \sim$ Po(np)   ($\lambda = np$)

Normal approximation to the Poisson distribution

When $\lambda > 15$,

    $X \sim$ Po(λ)

can be approximated by

    $X \sim$ N(λ, λ), with $\mu = \lambda$ and $\sigma = \sqrt{\lambda}$

A continuity correction must be applied.

Mixed Exercise 1

1 The random variable X has the distribution Po(6). Find:

(i) $P(X < 5)$

(ii) $P(6 < X \leqslant 9)$

(iii) the standard deviation of X.

2 At a certain factory the number of accidents per working day may be modelled by a Poisson distribution with parameter 0.5

(i) Find the probability that there are at least 2 accidents in a working day.

(ii) Find the probability that there are exactly 3 accidents in 2 working days.

(iii) In a period of 50 consecutive 5-day weeks, how many weeks would you expect to be accident-free?

3 People arrive randomly and independently at the elevator in a block of flats at an average rate of 4 people every 5 minutes.

(i) Find the probability that exactly two people arrive in a 1-minute period.

(ii) Find the probability that nobody arrives in a 15-second period.

(iii) The probability that at least one person arrives in the next t minutes is 0.9. Find the value of t.

Cambridge Paper 7 Q6 J08

4 The number of family groups arriving at a theme park may be modelled by a Poisson distribution with mean 1.5 per minute.

(i) Find the probability that during a particular 2-minute interval 3 family groups arrive at the theme park.

(ii) Find the probability that during a particular 5-minute interval more than 4 family groups arrive at the theme park.

(iii) Would the Poisson distribution provide a good model for the number of visitors to the theme park in a 1-minute interval? Give a reason for your answer.

5 2% of biscuits on a production line are broken. Broken biscuits occur randomly. 180 biscuits are checked to see whether they are broken.

Use a suitable approximation to find the probability that fewer than 4 are broken.

Cambridge Paper 7 Q1 N09(71)

6 The number of customers entering a certain branch of a bank may be modelled by a Poisson distribution with a mean of 2.4 per minute.

(i) Find the probability that 4 or more customers enter the branch during a 2-minute interval.

The probability is 0.002 that a customer who enters the branch intends to open a new account. During a particular morning 450 customers enter the bank.

(ii) Use a suitable approximation to find the probability that 3 or fewer of these 450 customers intend to open a new account.

7 In a certain country 0.3% of people have blood type AB⁻.

(i) Using a suitable approximation, find the probability that in a random sample of 1200 people from the country:

(a) at least 5 have blood type AB⁻

(b) fewer than 3 have blood type AB⁻.

(ii) A hospital urgently requires blood of type AB⁻. Assuming that blood donors are representative of the population as a whole, how large a random sample of donors must be taken in order that the probability of finding at least 1 donor with blood type AB⁻ is at least 99%?

8 Of people who wear contact lenses, 1 in 1500 on average have laser treatment for short sight.

(i) Use a suitable approximation to find the probability that, of a random sample of 2700 contact lens wearers, more than 2 people have laser treatment.

(ii) In a random sample of n contact lens wearers the probability that no one has laser treatment is less than 0.01. Find the least possible value of n.

Cambridge Paper 7 Q5 N04

9 A dressmaker makes dresses for Easifit Fashions. Each dress requires 2.5 m^2 of material. Faults occur randomly in the material at an average rate of 4.8 per 20m^2.

   (i) Find the probability that a randomly chosen dress contains at least 2 faults.

Each dress has a belt attached to it to make an outfit. Independently of faults in the material, the probability that a belt is faulty is 0.03. Find the probability that, in an outfit,

   (ii) neither the dress nor its belt is faulty,

   (iii) the dress has at least one fault and its belt is faulty.

The dressmaker attaches 300 randomly chosen belts to 300 randomly chosen dresses. An outfit in which the dress has at least one fault and its belt is faulty is rejected.

   (iv) Use a suitable approximation to find the probability that fewer than 3 outfits are rejected.

   Cambridge Paper 7 Q6 J06

10 At a particular shop, the number of people in a 20-minute period who buy the Daily Press is modelled by a random variable X with the distribution Po(3.8)

   (i) State two assumptions required for the Poisson model to be valid.

   (ii) Find the probability that

      (a) at least 1 person buys the Daily Press in a 20-minute period,

      (b) fewer than 4 people buy the Daily Press in a 15 minute period.

   (iii) The shop is open for 10 hours each day. At the beginning of the day the shop has 135 copies of the Daily Press to sell. Use a suitable approximation to find the probability that this will not be enough to satisfy the demand.

11 In a competition, the probability of buying a winning ticket is 0.002. Debbie sells 1000 tickets for the competition.

   (i) Using a Poisson approximation, find the probability that the number of winning tickets sold by Debbie is

      (a) fewer than three,

      (b) more than five.

   (ii) What is the minimum number of tickets Debbie must sell to have a 95% chance of selling at least one winning ticket?

2 Linear combinations of random variables

In this chapter you will learn about

- linear functions of a random variable
  - the mean and variance of $aX + b$
- linear combinations of random variables
  - the mean of $aX + bY$
  - the variance of $aX + bY$ when X and Y are independent
- the distribution of $X + Y$ when X and Y have independent Poisson distributions
- the distribution of $aX + b$ when X has a normal distribution
- the distribution of $aX + bY$ when X and Y have independent normal distributions

LINEAR FUNCTIONS OF A RANDOM VARIABLE

Consider the random variable X. You will recall from S1 that the mean or expectation of X is written $E(X)$ and denoted by μ and the variance of X is written $\text{Var}(X)$ and denoted by σ^2.

If you form a new random variable $aX + b$, where a and b are constants, this is known as a **linear function** of X.

Examples are

$$5X + 1, \quad \tfrac{1}{2}X - 7, \quad 6 - 2X, \quad 4X.$$

Note that a can be positive or negative and b can be positive, negative or zero.

Mean and variance of $aX + b$

The mean and variance of $aX + b$ can be found directly from $E(X)$ and $\text{Var}(X)$ as follows:

For any random variable X and constants a and b:

Mean

$$E(aX) = aE(X)$$

$$E(aX + b) = aE(X) + b$$

Variance

$$\text{Var}(aX) = a^2 \text{Var}(X)$$

$$\text{Var}(aX + b) = a^2 \text{Var}(X)$$

The variance is not affected by '$+ b$'.

↑

Notice a^2 here.

Example 2.1

The random variable X has mean 10 and variance 3.

(i) Find the mean and variance of the following random variables:

    (a) $W = 5X + 1$ (b) $V = 6 - 2X$.

(ii) Find the standard deviation of the random variable $Y = \tfrac{1}{2}X - 7$

You are given that $E(X) = 10$ and $Var(X) = 3$

  (i) (a) $W = 5X + 1$ $a = 5, b = 1$

           $E(W) = 5E(X) + 1 = 5 \times 10 + 1 = 51$

           $Var(W) = 5^2 Var(X) = 25 Var(X) = 25 \times 3 = 75$

     (b) $V = 6 - 2X$ $a = -2, b = 6$

           $E(V) = 6 - 2E(X) = 6 - 2 \times 10 = -14$

           $Var(V) = (-2)^2 Var(X) = 4 Var(X) = 4 \times 3 = 12$

 (ii) $Y = \frac{1}{2}X - 7$ $a = \frac{1}{2}, b = -7$

    First find the variance of Y.

    $Var(Y) = \left(\frac{1}{2}\right)^2 Var(X) = \frac{1}{4} \times 3 = 0.75$

    Now find the square root of the variance.

    s.d. of $Y = \sqrt{0.75} = 0.866$ (3 s.f.)

Example 2.2

A fair coin is tossed 10 times and the number of tails recorded.

  (i) The random variable X is the number of tails. State the mean and variance of X.

 (ii) The number of tails is multiplied by 3 and denoted by the random variable Y. State the mean and standard deviation of Y.

  (i) X is the number of tails in 10 tosses.

        $X \sim B(10, 0.5)$, with $n = 10$, $p = 0.5$, $q = 1 - p = 0.5$ Recall: Binomial distribution S1 Chapter 5.

    Mean $= E(X) = np = 10 \times 0.5 = 5$

    Variance $= npq = 10 \times 0.5 \times 0.5 = 2.5$

 (ii) You are given that $Y = 3X$. $a = 3, b = 0$

    $E(Y) = 3E(X) = 3 \times 5 = 15$

    $Var(Y) = 3^2 Var(X) = 9 Var(X) = 9 \times 2.5 = 22.5$

    s.d. of $Y = \sqrt{22.5} = 4.743... = 4.74$ (3 s.f.)

Example 2.3

Exam marks, X, have mean 70 and standard deviation 8.7. The marks need to be scaled using the formula $Y = aX + b$ so that the scaled marks, Y, have mean 55 and standard deviation 6.96. Find the values of a and b.

Cambridge Paper 7 Q1 J05

If $Y = aX + b$, then $E(Y) = aE(X) + b = a \times 70 + b$

But you are given that $E(Y) = 55$,

so $\quad 70a + b = 55 \qquad\qquad (1)$

$\text{Var}(Y) = a^2\text{Var}(X) = a^2 \times 8.7^2 = 75.69a^2$

But you are given that $\text{Var}(Y) = 6.96^2 = 48.4416$,

so $\quad 75.69a^2 = 48.4416$

$$a^2 = \frac{48.4416}{75.69} = 0.64$$

$$a = \sqrt{0.64} = 0.8$$

Alternatively

$a^2 \times 8.7^2 = 6.96^2$

$a \times 8.7 = 6.96$

$a = 0.8$

Substitute into (1):

$$70 \times 0.8 + b = 55$$

$$56 + b = 55$$

$$b = -1$$

Note that the formula for the scaling is $Y = 0.8X - 1$.

So $a = 0.8$ and $b = -1$.

LINEAR COMBINATIONS OF RANDOM VARIABLES

Now consider two random variables, X and Y. If you form a new random variable $aX + bY$, where a and b are constants, this is known as a **linear combination** of X and Y.

Examples are

$3X + 2Y$, $5X - 4Y$, $\frac{1}{2}X + \frac{3}{4}Y$.

Mean and variance of $aX \pm bY$

You can find the mean and variance of the new variable as follows:

For two random variables X and Y, and constants a and b,

$E(aX + bY) = aE(X) + bE(Y)$

$E(aX - bY) = aE(X) - bE(Y)$

If X and Y are **independent**,

$\text{Var}(aX + bY) = a^2\text{Var}(X) + b^2\text{Var}(Y)$

$\text{Var}(aX - bY) = a^2\text{Var}(X) + b^2\text{Var}(Y)$

The + sign here is important. Variances are <u>always</u> added.

Example 2.4

The random variables X and Y are such that $E(X) = 10$ and $E(Y) = 8$

Find:

(i) $E(X + Y)$ 　　　(ii) $E(X - Y)$ 　　　(iii) $E(3X + 2Y)$ 　　　(vi) $E\left(\frac{1}{2}X - \frac{3}{4}Y\right)$.

(i) $E(X + Y) = E(X) + E(Y) = 10 + 8 = 18$

(ii) $E(X - Y) = E(X) - E(Y) = 10 - 8 = 2$

(iii) $E(3X + 2Y) = 3E(X) + 2E(Y) = 3 \times 10 + 2 \times 8 = 46$

(iv) $E\left(\frac{1}{2}X - \frac{3}{4}Y\right) = \frac{1}{2}E(X) - \frac{3}{4}E(Y) = \frac{1}{2} \times 10 - \frac{3}{4} \times 8 = -1$

Example 2.5

X and Y are independent variables such that $\text{Var}(X) = 2$ and $\text{Var}(Y) = 3$

Find:

   (i) $\text{Var}(X + Y)$ (ii) $\text{Var}(X - Y)$ (iii) $\text{Var}(3X + 2Y)$ (iv) $\text{Var}\left(Y - \frac{1}{2}X\right)$.

   (i) $\text{Var}(X + Y) = \text{Var}(X) + \text{Var}(Y) = 2 + 3 = 5$

   (ii) $\text{Var}(X - Y) = \text{Var}(X) + \text{Var}(Y) = 2 + 3 = 5$ Notice that $\text{Var}(X + Y) = \text{Var}(X - Y)$.

Remember the + sign here.

   (iii) $\text{Var}(3X + 2Y) = 3^2\text{Var}(X) + 2^2\text{Var}(Y)$

$\qquad\qquad\qquad = 9\text{Var}(X) + 4\text{Var}(Y)$

$\qquad\qquad\qquad = 9 \times 2 + 4 \times 3$

$\qquad\qquad\qquad = 30$

   (iv) $\text{Var}\left(Y - \frac{1}{2}X\right) = \text{Var}(Y) + \left(\frac{1}{2}\right)^2\text{Var}(X)$

$\qquad\qquad\qquad = 3 + \frac{1}{4}\text{Var}(X)$

$\qquad\qquad\qquad = 3 + \frac{1}{4} \times 2$

$\qquad\qquad\qquad = 3.5$

Example 2.6

The independent random variables X and Y are such that X has mean 8 and variance 4.8 and Y has a Poisson distribution with mean 6. Find:

   (i) $E(2X - 3Y)$,   (ii) $\text{Var}(2X - 3Y)$.

Cambridge Paper 7 Q3 J04

$E(X) = 8$ and $\text{Var}(X) = 4.8$

$Y \sim \text{Po}(6)$, so $E(Y) = 6$ and $\text{Var}(Y) = 6$

   (i) $E(2X - 3Y) = 2E(X) - 3E(Y)$

$\qquad\qquad = 2 \times 8 - 3 \times 6$

$\qquad\qquad = -2$

(ii) $\text{Var}(2X - 3Y) = 2^2\text{Var}(X) + 3^2\text{Var}(Y)$

$$= 4\text{Var}(X) + 9\text{Var}(Y)$$

$$= 4 \times 4.8 + 9 \times 6$$

$$= 73.2$$

The results can be extended to any number of variables, as in the following example.

Example 2.7

Every morning Sophie's journey to work consists of three parts. She drives by car to the railway station, catches a train and then walks to her office. The times taken for each part of the journey are independent of each other.

The mean and standard deviation of the times taken for each part of her journey are shown in the table opposite.

(i)   Find the mean time for the complete journey.

(ii)   Find the standard deviation for the complete journey.

Part of journey	Mean time (min)	Standard deviation (min)
Car (C)	15	4
Train (R)	25	5
Walk (W)	10	3

Let T be the total time, where $T = C + R + W$

(i) $\text{E}(T) = \text{E}(C) + \text{E}(R) + \text{E}(W)$

$$= 15 + 25 + 10$$

$$= 50$$

The mean time for the complete journey is 50 min.

(ii)   *You must work in variances first.*

$$\text{Var}(T) = \text{Var}(C) + \text{Var}(R) + \text{Var}(W)$$

$$= 4^2 + 5^2 + 3^2$$

$$= 50$$

Now find the square root of the variance.

s.d. of $T = \sqrt{50} = 7.071\ldots$

The standard deviation of the time for the complete journey is 7.07 min (3 s.f.).

Independent observations X_1, X_2, ... of the random variable X

Special care must be taken when forming a new variable from independent observations of the **same** random variable.

Consider the random variable X with $\text{E}(X) = \mu$ and $\text{Var}(X) = \sigma^2$.

If you make two observations X_1 and X_2 from X and form a new variable $X_1 + X_2$, then

$$\text{E}(X_1 + X_2) = \text{E}(X_1) + \text{E}(X_2) = \mu + \mu = 2\mu = 2\text{E}(X)$$

If the observations are independent, then

$$\text{Var}(X_1 + X_2) = \text{Var}(X_1) + \text{Var}(X_2) = \sigma^2 + \sigma^2 = 2\sigma^2 = 2\text{Var}(X)$$

These results can be extended to n observations as follows:

For n observations of the random variable X,

$$\text{E}(X_1 + X_2 + \ldots + X_n) = n\text{E}(X)$$

If the observations are independent,

$$\text{Var}(X_1 + X_2 + \ldots + X_n) = n\text{Var}(X)$$

Example 2.8

Muffins are produced by a bakery. The weight, in grams, of a muffin is denoted by the random variable X. The mean of X is 50 g and the standard deviation is 3 g. The weight of a given muffin is independent of the weight of any other muffin.

The muffins are packed into boxes containing 10 muffins. Find the mean and standard deviation of the weight of the muffins in a box.

X is the weight, in grams, of a muffin. $\text{E}(X) = 50$ and $\text{Var}(X) = 3^2 = 9$

Let W be the weight, in grams, of 10 muffins. Then $W = X_1 + X_2 + X_3 + \ldots + X_{10}$

$$
\begin{aligned}
\text{E}(W) &= \text{E}(X_1) + \text{E}(X_2) + \text{E}(X_3) + \ldots + \text{E}(X_{10}) \\
&= 10\text{E}(X) \\
&= 10 \times 50 \\
&= 500
\end{aligned}
$$

As $X_1, X_2, \ldots$ are all taken from the same distribution, $\text{E}(X_1) = \text{E}(X)$, $\text{E}(X_2) = \text{E}(X)$ and so on.

The mean weight of muffins in a box is 500 g.

$$
\begin{aligned}
\text{Var}(W) &= \text{Var}(X_1) + \text{Var}(X_2) + \text{Var}(X_3) + \ldots + \text{Var}(X_{10}) \\
&= 10\text{Var}(X) \\
&= 10 \times 9 \\
&= 90
\end{aligned}
$$

$\text{Var}(X_1) = \text{Var}(X)$, $\text{Var}(X_2) = \text{Var}(X)$ and so on.

$$\text{s.d of } W = \sqrt{90} = 9.486\ldots$$

The standard deviation of the weight of muffins in a box is 9.49 g (3 s.f.).

Comparing the distributions of 2*X* and *X*₁ + *X*₂

Confusion sometimes arises between the random variable $2X$, a **multiple** of a single observation of X, and the random variable $X_1 + X_2$, the **sum** of two independent observations of X. You will see from the following example that their distributions are different.

Recall: Discrete random variables S1 (Chapter 4).

A fair tetrahedral die is thrown. Let X be the number on which it lands. The probability distribution of X is shown in the table.

x	1	2	3	4
$P(X = x)$	$\frac{1}{4}$	$\frac{1}{4}$	$\frac{1}{4}$	$\frac{1}{4}$

$$\text{E}(X) = 2.5 \text{ (by symmetry)}$$

$$\text{Var}(X) = \sum x^2 p - (\text{E}(X))^2$$

You could use $\text{E}(X) = \sum xp$ (S1 Chapter 4)

$$= 1^2 \times \tfrac{1}{4} + 2^2 \times \tfrac{1}{4} + 3^2 \times \tfrac{1}{4} + 4^2 \times \tfrac{1}{4} - 2.5^2$$

$$= 1.25$$

The variable 2X

Throw the die <u>once</u>.

Double the number on which the die lands and denote this by D, where $D = 2X$. $2X$ is a **multiple**.

The probability distribution and vertical line graph for D is as follows:

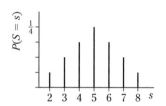

d	2	4	6	8
$P(D = d)$	$\frac{1}{4}$	$\frac{1}{4}$	$\frac{1}{4}$	$\frac{1}{4}$

$E(D) = 5$ (by symmetry)

$Var(D) = 2^2 \times \frac{1}{4} + 4^2 \times \frac{1}{4} + 6^2 \times \frac{1}{4} + 8^2 \times \frac{1}{4} - 5^2 = 5$

So, as expected,

$$E(D) = E(2X) = 2E(X)$$

$$Var(D) = Var(2X) = 2^2 Var(X) = 4Var(X)$$

The variable $X_1 + X_2$

Throw the die <u>twice</u>.

Let S be the sum of the numbers on which it lands, where $S = X_1 + X_2$.

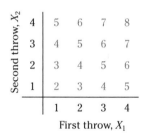

$X_1 + X_2$ is a **sum**.

The probability distribution of S is shown in the table and illustrated by the vertical line graph.

s	2	3	4	5	6	7	8
$P(S = s)$	$\frac{1}{16}$	$\frac{2}{16}$	$\frac{3}{16}$	$\frac{4}{16}$	$\frac{3}{16}$	$\frac{2}{16}$	$\frac{1}{16}$

By symmetry, $E(S) = 5$

$$Var(S) = \left(2^2 \times \frac{1}{16} + 3^2 \times \frac{2}{16} + \ldots + 8^2 \times \frac{1}{16}\right) - 5^2 = 2.5$$

So, as expected,

$$E(S) = E(X_1 + X_2) = 2E(X)$$

$$Var(S) = Var(X_1 + X_2) = 2Var(X)$$

Notice that the means of the two distributions are the same, but the variances are not. The random variable 'double the number' has a larger variance than the random variable 'the sum of the two numbers'.

Take care to distinguish between situations when a single observation has been multiplied by a constant (this is a **multiple**, as in $2X$, $3X$, $10X$) and when several different observations of the same random variable are added (this is a **sum**, as in $X_1 + X_2$, $X_1 + X_2 + X_3$, $X_1 + X_2 + \ldots + X_{10}$).

The general results are summarised below.

	Multiple	Sum
Mean	$E(nX) = nE(X)$	$E(X_1 + X_2 + \ldots + X_n) = nE(X)$
Variance	$Var(nX) = n^2 Var(X)$	$Var(X_1 + X_2 + \ldots + X_n) = nVar(X)$

Exercise 2a

1 The random variable X has mean 20 and variance 4. Find the mean and variance of the following random variables:

  (i) $Y = 6X + 1$ (ii) $D = 3X - 2$

  (iii) $W = \frac{1}{2}X - 4$ (iv) $V = -4X$.

2 (i) The random variable X has mean 12. The random variable $aX + 2$ has mean 5, where a is a constant. Find the value of a.

  (ii) The random variable Y has standard deviation 4. The random variable $aY - 1$ has standard deviation 40, where a is a constant. Find the value of a.

3 Test marks, X, have mean 60 and standard deviation 5. The marks are scaled using the formula $Y = aX + b$ so that the scaled marks, Y, have mean 80 and standard deviation 7.5

Find the values of a and b.

4 X and Y are independent random variables such that

$E(X) = 8$, $\mathrm{Var}(X) = 2$, $E(Y) = 10$, $\mathrm{Var}(Y) = 3$
Find:

  (i) $E(X + Y)$ (ii) $\mathrm{Var}(X - Y)$

  (iii) $E(3X + Y)$ (iv) $\mathrm{Var}(2X - 8)$

  (v) $E(2Y - \frac{1}{2}X + 5)$

  (vi) the standard deviation of $4X - 3Y + 5$.

5 The random variable X is distributed $B(20, 0.4)$. Find the mean and variance of the random variable Y, where $Y = 4X + 2$

6 The random variable W is distributed $\mathrm{Po}(16)$. Find the mean and standard deviation of the random variable V, where

$V = 3W - 2$

7 The independent Poisson random variables X and Y have parameters 2 and 5, respectively.

  (i) Find the mean and variance of the following random variables:

  (a) $X - Y$ (b) $3Y + 10$.

  (ii) For each of these random variables give **one** reason why the distribution is **not** Poisson.

8 The random variable X has mean 20 and standard deviation 3.

  (i) Find the mean and standard deviation of $4X$.

X_1, X_2, X_3 and X_4 are independent observations of X.

  (ii) Find the mean and standard deviation of $X_1 + X_2 + X_3 + X_4$

9 The random variable X has mean 20 and variance 4.

The variable Y is the sum of 5 independent observations of X. Find the mean and variance of Y.

10 A machine produces two types of items, A and B. The time taken to produce item A is distributed with mean 3 min and standard deviation 0.3 min. The time taken to produce item B is distributed with mean 5 min and standard deviation 0.9 min.

Find the mean and standard deviation of the time taken to produce:

  (i) 1 item of each type

  (ii) 4 items of Type A

  (iii) 4 items of Type A and 2 items of Type B.

11 In a certain city, the distribution of men's weights has a mean of 75 kg and a standard deviation of 5 kg. Find the mean and standard deviation of the distribution of the total weight of 4 randomly chosen men from the city.

THE SUM OF INDEPENDENT POISSON VARIABLES

The **sum** of two independent Poisson variables X and Y is also a Poisson variable. If the mean of X is λ_1 and the mean of Y is λ_2, then the mean of $X + Y$ is $\lambda_1 + \lambda_2$.

For two **independent Poisson variables** X and Y, where $X \sim \mathrm{Po}(\lambda_1)$ and $Y \sim \mathrm{Po}(\lambda_2)$,

  $X + Y \sim \mathrm{Po}(\lambda_1 + \lambda_2)$

  Mean $= \lambda_1 + \lambda_2$ Variance $= \lambda_1 + \lambda_2$

Example 2.9

The number of radioactive particles emitted per second by a certain metal is random and has mean 1.7. The radioactive metal is placed next to an object which independently emits particles at random such that the mean number of particles emitted per second is 0.6. Find the probability that the total number of particles emitted in the next 3 seconds is 6, 7 or 8.

Cambridge Paper 7 Q1 N04

Let X be the number of radioactive particles emitted by the metal in 3 seconds.

> In 1 second, mean = 1.7
> In 3 seconds, mean = $3 \times 1.7 = 5.1$

So $X \sim \text{Po}(5.1)$

Let Y be the number of radioactive particles emitted by the object in 3 seconds.

> In 1 second, mean = 0.6
> In 3 seconds, mean = $3 \times 0.6 = 1.8$

So $Y \sim \text{Po}(1.8)$

Let T be the total number of radioactive particles emitted in 3 seconds, where $T = X + Y$

$T \sim \text{Po}(5.1 + 1.8)$, i.e. $T \sim \text{Po}(6.9)$

$\text{P}(T = 6, 7 \text{ or } 8)$

$$= \text{P}(T = 6) + \text{P}(T = 7) + \text{P}(T = 8)$$

$$= e^{-6.9}\left(\frac{6.9^6}{6!} + \frac{6.9^7}{7!} + \frac{6.9^8}{8!}\right)$$

$$= 0.4283\ldots$$

$$= 0.428 \text{ (3 s.f.)}$$

Example 2.10

During a weekday, heavy lorries pass a checkpoint on a village street independently and at random times. The mean rate for the westbound lorries is 6 in any 30-minute period and, independently, the mean rate for eastbound lorries is 3 in any 30-minute period.

Find the probability that:

(i)   no lorries pass the checkpoint in a given 10-minute interval

(ii) exactly four lorries pass the checkpoint in a given 20-minute interval

(iii) at least one lorry from each direction will pass the checkpoint in a given 20-minute interval.

(i) Let X be the number of westbound lorries in 10 min, where $X \sim \text{Po}(2)$

Let Y be the number of eastbound lorries in 10 min, where $Y \sim \text{Po}(1)$

Let T be the total number of lorries in 10 min, where $T = X + Y$

So $T \sim \text{Po}(2 + 1)$, i.e. $T \sim \text{Po}(3)$

$\text{P}(T = 0) = e^{-3} = 0.04978\ldots = 0.0498 \text{ (3 s.f.)}$

(ii) Let S be the total number of lorries in 20 min, where $S \sim \text{Po}(6)$ Define a new variable.

$$P(S = 4) = e^{-6} \times \frac{6^4}{4!} = 0.1338\ldots = 0.134 \text{ (3 s.f.)}$$

(iii) Let U be number of lorries westbound in 20 min, where $U \sim \text{Po}(4)$ Define new variables.

$$P(U \geqslant 1) = 1 - P(U = 0) = 1 - e^{-4}$$

Let V be the number of lorries eastbound in 20 min, where $V \sim \text{Po}(2)$

$$P(V \geqslant 1) = 1 - P(V = 0) = 1 - e^{-2}$$

P(at least 1 lorry from each direction)

$$= P(U \geqslant 1) \times P(V \geqslant 1)$$

$$= (1 - e^{-4}) \times (1 - e^{-2})$$

$$= 0.8488\ldots$$

$$= 0.849 \text{ (3 s.f.)}$$

Example 2.11

In restaurant A an average of 2.2% of tablecloths are stained and, independently, in restaurant B an average of 5.8% of tablecloths are stained.

(i) Random samples of 55 tablecloths are taken from each restaurant. Use a suitable Poisson approximation to find the probability that a total of more than 2 tablecloths are stained.

(ii) Random samples of n tablecloths are taken from each restaurant. The probability that at least one tablecloth is stained is greater than 0.99. Find the least possible value of n.

Cambridge Paper 7 Q6 J10(71)

(i) Let X be the number of stained tablecloths in restaurant A,
where $X \sim \text{B}(55, 0.022)$ with $n = 55$ and $p = 0.022$

$$np = 55 \times 0.022 = 1.21$$

Since $n > 50$ and $np < 5$, use a Poisson approximation, where $X \sim \text{Po}(1.21)$

Let Y be the number of stained tablecloths in restaurant B,
where $Y \sim \text{B}(55, 0.058)$ with $n = 55$ and $p = 0.058$

$$np = 55 \times 0.058 = 3.19$$

Since $n > 50$ and $np < 5$, use a Poisson approximation, where $Y \sim \text{Po}(3.19)$

Let T be the total number of stains, where $T = X + Y$

$T \sim \text{Po}(1.21 + 3.19)$, i.e. $T \sim \text{Po}(4.4)$

$$P(T > 2) = 1 - P(T \leqslant 2)$$

$$= 1 - e^{-4.4} \left(1 + 4.4 + \frac{4.4^2}{2!} \right)$$

$$= 0.8148\ldots$$

$$= 0.815 \text{ (3 s.f.)}$$

(ii) When n tablecloths are taken, assuming $n > 50$ and $np < 5$,

$$T \sim \text{Po}(0.022n + 0.058n), \text{ i.e. } T \sim \text{Po}(0.08n)$$

P(at least 1 tablecloth is stained)

$$= 1 - \text{P(no tablecloths are stained)}$$

$$= 1 - e^{-0.08n}$$

You require $1 - e^{-0.08n} > 0.99$

$$0.01 > e^{-0.08n}$$

$$e^{-0.08n} < 0.01$$

Take $\log_e$ of both sides:

$$-0.08n < \ln(0.01)$$

Divide by $(-0.08n)$.   Remember to reverse the inequality.

$$n > \frac{\ln(0.01)}{-0.08}$$

$$n > 57.56\ldots$$

So the least possible value of n is 58.

Exercise 2b

1 X and Y are independent random variables, where

   $X \sim \text{Po}(2)$ and $Y \sim \text{Po}(3.5)$

   Find:

   (i) $\text{P}(X + Y = 6)$

   (ii) $\text{P}(X + Y < 3)$

   (iii) $\text{P}(X + Y \geqslant 4)$.

2 During the evening, vehicles pass a certain marker point on a motorway at random times at a constant average rate of 2 vehicles every 3 minutes.

   (i) Find the probability that fewer than 4 vehicles pass the marker point in a period of 3 minutes.

   (ii) Find the probability that the total number of vehicles passing the marker point in two separate periods of 3 minutes and 5 minutes is 10.

3 A restaurant kitchen has two microwave ovens, A and B. The number of times that A breaks down has a Poisson distribution with mean 0.4 per week. Independently of A, the number of times that B breaks down has a

Poisson distribution with mean 0.1 per week. Find the probability that, in the next 3 weeks:

   (i) A will not break down and B will break down twice

   (ii) there will be a total of 3 breakdowns.

4 Two identical racing cars are being tested independently on a circuit. For each car, the number of mechanical breakdowns may be modelled by a Poisson distribution with a mean of 1 breakdown in 100 laps. If a car breaks down, the service team attends to it and the car then continues on the circuit. The first car completes 20 laps and the second car completes 40 laps. Find the probability that, during the test drives, the service team is called out to attend to breakdowns:

   (i) once

   (ii) more than twice.

5 The number of detached houses sold by an estate agent may be modelled by a Poisson distribution with mean 2.75 per week and, independently, the number of semi-detached houses sold may be modelled by a Poisson distribution with mean 3.25

(i) Find the mean and variance of the total number of houses sold in a week.

(ii) Find the probability that fewer than 5 houses are sold in a week.

(iii) Using a suitable approximation, find the probability that more than 30 houses are sold in a 4-week period.

6 In hockey matches, the school team scores goals independently and at random times. In home games they score an average of 3.1 goals per match and in away games they score an average of 2.5 goals per match.

(i) Find the probability that the team will score a total of 8 goals in their next 2 home games.

The team captain sets a target of a total of more than 7 goals in the next 2 games consisting of a home game followed by an away game.

(ii) Find the probability that the team will meet the target.

7 The centre pages of the *Weekly Press* consist of a page of film reviews and a page of classified advertisements. The number of misprints on the film-reviews page may be modelled by a Poisson distribution with mean 1.4 and, independently, the number of misprints on the classified-advertisement page may be modelled by a Poisson distribution with mean 2.1

(i) Find the probability that there will be more than 4 misprints on the centre pages of the *Weekly Press*.

(ii) Find the least integer n such that the probability that there are more than n misprints on the centre pages is less than 10%.

8 People arrive at a checkout at an average rate of 3 people every 4 minutes.

(i) State two conditions that must be satisfied for the number of people arriving to follow a Poisson distribution.

(ii) Assuming that these conditions are satisfied, find the probability that exactly 4 people will arrive in a 5-minute period.

At another checkout the number of people arriving per minute has the distribution Po(1).

(iii) Find the probability that a total of fewer than 3 people arrive at the two checkouts in a 6-minute period.

THE SUM OF INDEPENDENT NORMAL VARIABLES

Consider two independent normal variables, X and Y, where $X \sim \mathrm{N}(\mu_1, \sigma_1^2)$ and $Y \sim \mathrm{N}(\mu_2, \sigma_2^2)$

If you form the new variable $X + Y$, then this is also **normally distributed**.

*$X + Y$ is the **sum** of the two variables.*

The mean of $X + Y$ is $\mu_1 + \mu_2$ and the variance is $\sigma_1^2 + \sigma_2^2$, so

$$X + Y \sim \mathrm{N}(\mu_1 + \mu_2, \sigma_1^2 + \sigma_2^2)$$

Example 2.12

A coffee machine is installed in a students' common room. It dispenses white coffee by first releasing a quantity of black coffee and then adding a quantity of milk. The amount of black coffee dispensed is normally distributed with mean 122.5 ml and standard deviation 7.5 ml and, independently, the amount of milk dispensed is normally distributed with mean 30 ml and standard deviation 5 ml.

Each cup is marked to a level of 137.5 ml. If this level is not attained, the customer receives the drink free of charge.

What percentage of cups of white coffee will be given free of charge?

Let B be the amount, in millilitres, of black coffee dispensed, where $B \sim N(122.5, 7.5^2)$

Let M be the amount, in millilitres, of milk dispensed, where $M \sim N(30, 5^2)$

Let W be the amount, in millilitres, of white coffee dispensed, made by combining the black coffee and the milk, where $W = B + M$

$E(W) = E(B) + E(M) = 122.5 + 30 = 152.5$

$Var(W) = Var(B) + Var(M) = 7.5^2 + 5^2 = 81.25$

Since B and M are independent normal variables, W is also a normal variable.

So $W \sim N(152.5, 81.25)$ with $\mu = 152.5$ and $\sigma = \sqrt{81.25}$ Do not calculate $\sqrt{81.25}$ yet as this could lead to inaccuracy.

The drink is free of charge if $W < 137.5$

$P(W < 137.5)$

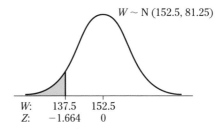

$W \sim N(152.5, 81.25)$

$\quad = P\left(Z < \dfrac{137.5 - 152.5}{\sqrt{81.25}}\right)$

$\quad = P(Z < -1.664)$

$\quad = 1 - \Phi(1.664)$

$\quad = 1 - 0.9519$

$\quad = 0.0481$

W:	137.5	152.5
Z:	-1.664	0

So, approximately 5% of the cups of white coffee will be given free of charge.

The result can be extended to any set of independent normal variables $X_1, X_2, \ldots, X_n$, where, with obvious notation,

$$X_1 + X_2 + \ldots + X_n \sim N(\mu_1 + \mu_2 + \ldots + \mu_n, \sigma_1^2 + \sigma_2^2 + \ldots + \sigma_n^2)$$

Example 2.13

Four runners, Andy, Bob, Chris and Dai, train to take part in a 1600 m relay race in which Andy is to run 100 m, Bob 200 m, Chris 500 m and Dai 800 m.

During training their individual times, recorded in seconds, follow normal distributions as follows:

	Andy	**Bob**	**Chris**	**Dai**
Mean	10.8	23.7	62.8	121.2
Standard deviation	0.2	0.3	0.9	2.1

Find the probability that they will run the relay race in less than 3 minutes 35 seconds.

With obvious notation,

$A \sim N(10.8, 0.2^2)$, $B \sim N(23.7, 0.3^2)$, $C \sim N(62.8, 0.9^2)$ and $D \sim N(121.2, 2.1^2)$

Let T be the total time, in seconds, for the relay race, where $T = A + B + C + D$

$E(T) = E(A) + E(B) + E(C) + E(D)$

$\quad = 10.8 + 23.7 + 62.8 + 121.2$

$\quad = 218.5$

$$\text{Var}(T) = \text{Var}(A) + \text{Var}(B) + \text{Var}(C) + \text{Var}(D)$$
$$= 0.2^2 + 0.3^2 + 0.9^2 + 2.1^2$$
$$= 5.35$$

So $T \sim N(218.5, 5.35)$, with $\mu = 218.5$ and $\sigma = \sqrt{5.35}$

3 minutes 35 seconds = 215 seconds

$$P(T < 215) = P\left(Z < \frac{215 - 218.5}{\sqrt{5.35}}\right)$$

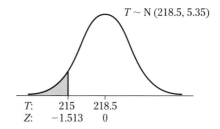

$T \sim N(218.5, 5.35)$

$$= P(Z < -1.513)$$
$$= 1 - \Phi(1.513)$$
$$= 1 - 0.9349$$
$$= 0.0651$$

T:	215	218.5
Z:	-1.513	0

The probability that the runners take less than 3 minutes 35 seconds is 0.0651

Independent observations from the same distribution

Now consider the special case when $X_1, X_2, ..., X_n$ are **independent observations** from the **same** normal distribution X, where $X \sim N(\mu, \sigma^2)$

Since they are all observations from X,

$$X_1 \sim N(\mu, \sigma^2)$$
$$X_2 \sim N(\mu, \sigma^2)$$
$$... \qquad ...$$
$$X_n \sim N(\mu, \sigma^2)$$

The sum $X_1 + X_2 + ... + X_n$ is also normally distributed.

Mean $E(X_1 + X_2 + ... + X_n) = \mu + \mu + ... + \mu = n\mu$

Variance $\text{Var}(X_1 + X_2 + ... + X_n) = \sigma^2 + \sigma^2 + ... + \sigma^2 = n\sigma^2$

If $X \sim N(\mu, \sigma^2)$, then the **sum of n independent observations** of X is also normally distributed,

where

$$X_1 + X_2 + ... + X_n \sim N(n\mu, n\sigma^2)$$

Example 2.14

A baker makes digestive biscuits and packs them into packets containing 12 biscuits. The weight of a biscuit is normally distributed with mean 20 g and standard deviation 2 g. Find the probability that the total weight of the biscuits in a packet is greater than 230 g.

Let B be the weight, in grams, of a biscuit, where $B \sim N(20, 2^2)$

Let T be the total weight, in grams, of 12 biscuits.

Then $T = B_1 + B_2 + \dots + B_{12}$

$E(T) = 12E(B) = 12 \times 20 = 240$

$\text{Var}(T) = 12\text{Var}(B) = 12 \times 2^2 = 48$ Note: $\text{Var}(T)$ is not equal to $12^2\text{Var}(B)$.

So $T \sim N(240, 48)$, with $\mu = 240$ and $\sigma = \sqrt{48}$

$$P(T > 230) = P\left(Z > \frac{230 - 240}{\sqrt{48}}\right)$$

$$= P(Z > -1.443)$$

$$= \Phi(1.443)$$

$$= 0.9255$$

$$= 0.926 \text{ (3 s.f.)}$$

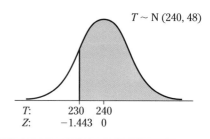

$T \sim N(240, 48)$

T:	230	240
Z:	-1.443	0

Example 2.15

The maximum load a lift can carry is 450 kg. In a certain country, the weights of men are normally distributed with mean 60 kg and standard deviation 10 kg. The weights of women are normally distributed with mean 55 kg and standard deviation 5 kg.

A total of 5 men and 2 women enter a lift. Find the probability that the lift will be overloaded, assuming that their weights are independent.

Let M be the weight, in kilograms, of a man, where $M \sim N(60, 10^2)$

Let W be the total weight, in kilograms, of a woman, where $W \sim N(55, 5^2)$

The lift is overloaded if

$$M_1 + M_2 + M_3 + M_4 + M_5 + W_1 + W_2 > 450$$

Let $T = M_1 + M_2 + M_3 + M_4 + M_5 + W_1 + W_2$

$$E(T) = 5E(M) + 2E(W)$$

$$= 5 \times 60 + 2 \times 55$$

$$= 410$$

$$\text{Var}(T) = 5\text{Var}(M) + 2\text{Var}(W)$$

$$= 5 \times 10^2 + 2 \times 5^2$$

$$= 550$$

Since M and W are normally distributed, T is also normally distributed, so $T \sim N(410, 550)$ with $\mu = 410$, $\sigma = \sqrt{550}$

$$P(T > 450) = P\left(Z > \frac{450 - 410}{\sqrt{550}}\right)$$

$$= P(Z > 1.706)$$

$$= 1 - \Phi(1.706)$$

$$= 1 - 0.9559$$

$$= 0.0441$$

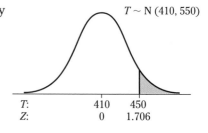

$T \sim N(410, 550)$

T:	410	450
Z:	0	1.706

The probability that the lift is overloaded is 0.0441

THE DIFFERENCE OF INDEPENDENT NORMAL VARIABLES

For two independent normal variables X and Y, where $X \sim N(\mu_1, \sigma_1^2)$ and $Y \sim N(\mu_2, \sigma_2^2)$, the variable $X - Y$ is also **normally distributed**.

The mean of $X - Y$ is $\mu_1 - \mu_2$ and the variance of $X - Y$ is $\sigma_1^2 + \sigma_2^2$,

so

$$X - Y \sim N(\mu_1 - \mu_2, \sigma_1^2 + \sigma_2^2)$$ $X - Y$ is the **difference** of the variables.

Remember the $+$ sign here.

Example 2.16

A certain liquid medicine is sold in bottles containing a nominal 20 ml of medicine. Tests on a large number of bottles indicate that the volume of liquid in a bottle follows a normal distribution with mean 20.42 ml and standard deviation 0.429 ml. The capacity of the bottles is normally distributed with mean 21.77 ml and standard deviation 0.210 ml.

Estimate the percentage of bottles that overflow during filling.

Let X be the volume, in millilitres, of liquid dispensed into a bottle, where $X \sim N(20.42, 0.429^2)$

Let Y be the capacity, in millilitres, of a bottle, where $Y \sim N(21.77, 0.210^2)$

The medicine will overflow if the volume of liquid is greater than the capacity of the bottle, i.e. if $X > Y$, so $X - Y > 0$.   Write the inequality with all the letters on one side.

Let $D = X - Y$

$$E(D) = E(X) - E(Y)$$

$$= 20.42 - 21.77$$

$$= -1.35$$

$$Var(D) = Var(X) + Var(Y)$$ Remember $+$ sign here.

$$= 0.429^2 + 0.210^2$$

$$= 0.228141$$

$D \sim N(-1.35, 0.228141)$ with $\mu = -1.35$ and $\sigma = \sqrt{0.228141}$

You require $P(X - Y > 0) = P(D > 0)$

$$P(D > 0) = P\left(Z > \frac{0 - (-1.35)}{\sqrt{0.228141}}\right)$$ $D \sim N(-1.35, 0.2281)$

$$= P(Z > 2.826)$$

$$= 1 - \Phi(2.826)$$

$$= 1 - 0.9976$$

D:	−1.35	0
Z:	0	2.826

$$= 0.0024$$

So 0.24% of bottles overflow during filling.

Example 2.17

A machine produces rubber balls whose diameters are normally distributed with mean 5.50 cm and standard deviation 0.08 cm. The balls are packed in cylindrical tubes whose internal diameters are normally distributed with mean 5.70 cm and standard deviation 0.12 cm.

A randomly selected ball is placed in a randomly selected tube. The clearance, C cm, is the internal diameter of the tube minus the diameter of the ball.

(i) Describe the distribution of C.

(ii) Find the probability that the clearance is between 0.05 cm and 0.25 cm.

(i) Let B be the diameter, in centimetres, of a rubber ball, where $B \sim N(5.50, 0.08^2)$

Let T be the internal diameter, in centimetres, of a cylindrical tube, where $T \sim N(5.70, 0.12^2)$

You are given that $C = T - B$

$$E(C) = E(T) - E(B)$$
$$= 5.70 - 5.50$$
$$= 0.2$$

$$Var(C) = Var(T) + Var(B) \qquad \text{Remember + sign here.}$$
$$= 0.08^2 + 0.12^2$$
$$= 0.0208$$

$C \sim N(0.2, 0.0208)$ with $\mu = 0.2$ and $\sigma = \sqrt{0.0208}$

(ii) $P(0.05 < C < 0.25)$

$$= P\left(\frac{0.05 - 0.2}{\sqrt{0.0208}} < Z < \frac{0.25 - 0.2}{\sqrt{0.0208}}\right)$$
$$= P(-1.040 < Z < 0.347)$$
$$= \Phi(0.347) - (1 - \Phi(1.040))$$
$$= 0.6357 - (1 - 0.8508)$$
$$= 0.4865$$

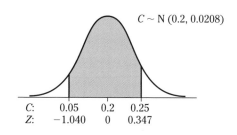

$C \sim N(0.2, 0.0208)$

| C: | 0.05 | 0.2 | 0.25 |
| Z: | -1.040 | 0 | 0.347 |

The probability that the clearance is between 0.05 cm and 0.25 cm is 0.487 (3 s.f.).

Example 2.18

A certain make of washing machine has a wash-time with mean 56.9 minutes and standard deviation 4.8 minutes. A certain make of tumble dryer has a drying-time with mean 61.1 minutes and standard deviation 6.3 minutes. Both times are normally distributed and are independent of each other. Find the probability that a randomly chosen wash-time differs by more than 3 minutes from a randomly chosen drying-time.

Cambridge Paper 7 Q4 J06

Let W be the washing time, in minutes, where $W \sim N(56.9, 4.8^2)$

Let D be the drying time, in minutes, where $D \sim N(61.1, 6.3^2)$

If W and D differ by more than 3 min, then $D - W > 3$ or $W - D > 3$

Now $W - D > 3$ can be rearranged as $D - W < -3$, so you need to find
$P(D - W > 3) + P(D - W < -3)$

Let $T = D - W$

$$E(T) = E(D) - E(W) = 61.1 - 56.9 = 4.2$$

$$\text{Var}(T) = \text{Var}(D) + \text{Var}(W) = 6.3^2 + 4.8^2 = 62.73$$

So $T \sim N(4.2, 62.73)$ with $\mu = 4.2$ and $\sigma = \sqrt{62.73}$

$$P(T > 3) = P\left(Z > \frac{3 - 4.2}{\sqrt{62.73}}\right)$$

$$= P(Z > -0.152)$$

$$= \Phi(0.152)$$

$$= 0.5604$$

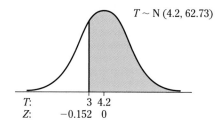

$$P(T < -3) = P\left(Z < \frac{-3 - 4.2}{\sqrt{62.73}}\right)$$

$$= P(Z < -0.909)$$

$$= 1 - \Phi(0.909)$$

$$= 1 - 0.8182$$

$$= 0.1818$$

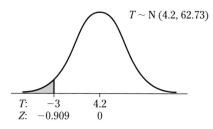

So P(washing time differs by more than 3 min from drying time)

$$= P(|T| > 3)$$

$$= P(T > 3) + P(T < -3) \qquad \text{If the difference is greater than 3,}$$
$$\text{then } |T| > 3, \text{ i.e. } T > 3 \text{ or } T < -3$$

$$= 0.5604 + 0.1818$$

$$= 0.7422$$

$$= 0.742 \text{ (3 s.f.)}$$

Special case when two observations are from the same distribution

Note the following *special case* when the two observations are from the **same** distribution. In this case the **mean of the differences** is **zero**.

Example 2.19

Bags of carrots have weights, in kilograms, that are normally distributed with mean 25.4 and variance 0.08. Find the probability that the difference in weight between two randomly selected bags of carrots is more than 1 kg.

Let X be the weight, in kilograms, of a bag of carrots, where $X \sim N(25.4, 0.08)$

Let $D = X_1 - X_2$

$$E(D) = 25.4 - 25.4 = 0$$

$$\text{Var}(D) = 0.08 + 0.08 = 0.16$$

$D \sim N(0, 0.16)$ with $\mu = 0$ and $\sigma = \sqrt{0.16} = 0.4$

You require P(|D| > 1) = P(D > 1) + P(D < −1)

$$P(D > 1) = P\left(Z > \frac{1 - 0}{0.4}\right)$$

$$= P(Z > 2.5)$$

$$= 1 - \Phi(2.5)$$

$$= 1 - 0.9938$$

$$= 0.0062$$

By symmetry, P(D < −1) = 0.0062

So P(|D| > 1) = 2 × 0.0062 = 0.0124

The probability that the difference in weight between two bags of carrots is more than 1 kg is 0.0124

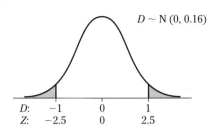

$D \sim N(0, 0.16)$

| D: | −1 | 0 | 1 |
| Z: | −2.5 | 0 | 2.5 |

Questions may involve both sums and differences.

Example 2.20

In a cafeteria, baked beans are served either as ordinary portions or children's portions. The amount given for an ordinary portion is a normal variable with mean 90 g and variance 9 g². Independently, the quantity given for a children's portion is a normal variable with mean 43 g and variance 4 g².

Tom has 2 children's portions of baked beans and his father has an ordinary portion. Find the probability that Tom is given more than his father.

Let A be the quantity, in grams, of an ordinary portion, where $A \sim N(90, 9)$

Let C be the quantity, in grams, of a children's portion, where $C \sim N(43, 4)$

Notice that the **variance** has been given, rather than the standard deviation.

You are required to find

$$P(C_1 + C_2 > A), \text{ i.e. } P(C_1 + C_2 - A > 0)$$

Write the inequality with all the letters on one side.

Let $W = C_1 + C_2 - A$

$$E(W) = 2E(C) - E(A)$$

$$= 2 \times 43 - 90$$

$$= -4$$

$$Var(W) = 2Var(C) + Var(A)$$

$$= 2 \times 4 + 9$$

$$= 17$$

So $W \sim N(-4, 17)$ with $\mu = -4$ and $\sigma = \sqrt{17}$

$$P(W > 0) = P\left(Z > \frac{0 - (-4)}{\sqrt{17}}\right)$$

$$= P(Z > 0.970)$$

$$= 1 - \Phi(0.970)$$

$$= 1 - 0.8340$$

$$= 0.1660$$

$W \sim N(-4, 17)$

| W: | −4 | 0 |
| Z: | 0 | 0.970 |

The probability that Tom has more than his father is 0.166 (3 s.f.).

Example 2.21

Weights of garden tables are normally distributed with mean 36 kg and standard deviation 1.6 kg. Weights of garden chairs are normally distributed with mean 7.3 kg and standard deviation 0.4 kg. Find the probability that the total weight of 2 randomly chosen tables is more than the total weight of 10 randomly chosen chairs.   Cambridge Paper 7 Q3 N08

Let T be the weight, in kilograms, of a table, where $T \sim N(36, 1.6^2)$

Let C be the weight, in kilograms, of a chair, where $C \sim N(7.3, 0.4^2)$

Now let $X = T_1 + T_2$

$\qquad E(X) = 2E(T) = 2 \times 36 = 72$

$\qquad Var(X) = 2Var(T) = 2 \times 1.6^2 = 5.12$

So $X \sim N(72, 5.12)$

Let $Y = C_1 + C_2 + \dots + C_{10}$

$\qquad E(Y) = 10E(C) = 10 \times 7.3 = 73$

$\qquad Var(Y) = 10 \times 0.4^2 = 1.6$

So $Y \sim N(73, 1.6)$

You require $P(X > Y)$, i.e. $P(X - Y > 0)$

Let $D = X - Y$

$\qquad E(D) = E(X) - E(Y) = 72 - 73 = -1$

$\qquad Var(D) = Var(X) + Var(Y) = 5.12 + 1.6 = 6.72$

$D \sim N(-1, 6.72)$ with $\mu = -1$ and $\sigma = \sqrt{6.72}$

$\qquad P(D > 0) = P\left(Z > \dfrac{0 - (-1)}{\sqrt{6.72}}\right)$

$\qquad\qquad\quad = P(Z > 0.386)$

$\qquad\qquad\quad = 1 - \Phi(0.386)$

$\qquad\qquad\quad = 1 - 0.6502$

$\qquad\qquad\quad = 0.3498$

> Tip:
> Take the time to define the variables carefully. This will help you to avoid errors.

> Rewrite the inequality so that the variables are on the same side.
>
> Define a new variable.

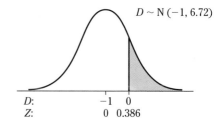

So the probability that the total weight of two tables is more than the total weight of 10 chairs is 0.350 (3 s.f.).

Exercise 2c

1 X and Y are independent random variables, where $X \sim N(20, 3^2)$ and $Y \sim N(30, 3.5^2)$

   (i) Find the mean and standard deviation of the distribution of $X + Y$.

  (ii) State the distribution of $X + Y$.

 (iii) Find: (a) $P(X + Y < 60)$
               (b) $P(38 < X + Y < 45)$.

2 At a health centre, consultations with a nurse take X minutes, where X may be modelled by a normal distribution with mean 5.4 and standard deviation 1.3. Consultations with a doctor take Y minutes, where Y may be modelled by a normal distribution with mean 9.2 and standard deviation 2.4.

Lesley has a consultation with a nurse, followed immediately by a consultation with a doctor.

Find the probability that the total duration of Lesley's consultations is:

(i) more than 20 min

(ii) less than 15 min

(iii) between 11.5 min and 14 min.

3 A manufacturer produces jars of honey. In the production process, the weight of honey dispensed into a jar is normally distributed with mean 345 g and standard deviation 3 g. The weight of an empty jar is normally distributed with mean 180 g and standard deviation 2 g and the weight of a jar lid is normally distributed with mean 5 g and standard deviation 0.5 g.

(i) Find the distribution of the total weight of a jar of honey (honey, jar and lid).

(ii) A jar of honey is selected from the jars ready for despatch. Find the probability that it weighs:

   (a) more than 525 g

   (b) between 532 g and 538 g.

4 The random variable X has the distribution $N(24.3, 3.5^2)$. The sum of 10 independent observations of X is denoted by Y.

(i) Find $E(Y)$ and $Var(Y)$.

(ii) Find $P(Y > 260.1)$.

(iii) Given that $P(Y < c) = 0.95$, find c.

5 The weight of a Chocolate Delight cake is normally distributed with mean 20 g and standard deviation 2 g. The cakes are sold in boxes containing 6 cakes. The weight of the box is normally distributed with mean 30 g and standard deviation 4 g.

(i) Find the probability that the total weight of 6 Chocolate Delight cakes is less than 110 g.

(ii) Find the probability that the total weight of a box containing 6 cakes is:

   (a) more than 162 g

   (b) less that 137 g

   (c) between 140 g and 153 g.

6 In a physics laboratory there are two types of resistors, A and B.

The resistance of Type A may be modelled by a normal variable with mean 4 ohms and standard deviation 0.12 ohms.

The resistance of Type B can be modelled by a normal variable with mean 2 ohms and standard deviation 0.05 ohms.

(i) Matt selects a resistor of each type at random and connects them to make a 6-ohm resistor. Find the probability that it has a resistance greater than 6.25 ohms.

(ii) Molly selects 3 Type B resistors at random and connects them to make a 6-ohm resistor. Find the probability that it has a resistance greater than 6.25 ohms.

7 X and Y are independent random variables where $X \sim N(110, 576)$ and $Y \sim N(100, 49)$

(i) Find the mean and the standard deviation of the distribution of $X - Y$.

(ii) State the distribution of $X - Y$.

(iii) Find $P(X - Y > 0)$.

(iv) Find $P(X - Y < 15)$.

8 Bolts are manufactured which are to fit in holes in steel plates. The diameter of the bolts is normally distributed with mean 2.60 cm and standard deviation 0.03 cm. The diameter of the holes is normally distributed with mean 2.71 cm and standard deviation 0.04 cm.

(i) Show that, if a bolt and a hole are selected at random, the probability that the bolt is too large to enter the hole is 0.0139

(ii) The random selection of a bolt and hole is carried out 5 times. Find the probability that in every case the bolt will be able to enter the hole.

9 In a certain village, the heights of the women are normally distributed with mean 164 cm and standard deviation 5 cm. The heights of the men are normally distributed with mean 173 cm and standard deviation 6 cm.

A man and a woman are selected at random from the adults in the village. Find the probability that:

(i) the woman is taller than the man

(ii) the man is more than 5 cm taller than the woman

(iii) the difference between the heights of the man and the woman is more than 5 cm.

10 A hardware store sells planks of wood. The lengths, in centimetres, of the planks are normally distributed with mean 210 and standard deviation 7.

Two planks are selected at random. Find the probability that they differ in length by more than 10 cm.

11 Each weekday Mr Harper, a retired statistician, walks to his local library to read the newspapers and then walks back home. He proposes the following model.

The time he spends travelling is normally distributed with mean 15 minutes and standard deviation 2 minutes.

The time he spends in the library is normally distributed with mean 25 minutes and standard deviation 4 minutes.

Assume that the model is true.

(i) Mr Harper sets out to walk to the library at 11.45am. His wife prepares lunch for 12.30pm. Find the probability that he is late for lunch.

(ii) Find the probability that, on a particular day, Mr Harper spends more time travelling than he spends in the library.

12 Fiona, Carly, Jenny and Vicky are in the 4×100 m freestyle relay team, each swimming 100 m.

Their individual times to swim 100 m, recorded in seconds, may be modelled by normal variables with the following parameters:

	Fiona	Carly	Jenny	Vicky
Mean (s)	52.5	52.0	53.5	51.5
Standard deviation (s)	0.3	0.6	1.2	0.6

Calculate the probability that, in a particular race:

(i) Fiona will swim her leg in less than 52.5 seconds

(ii) the relay team will take less than 3 minutes 31.3 seconds to swim the race

(iii) Carly will swim her leg faster than Vicky.

13 Rods are produced in two lengths, 'short' and 'long'.

S is the length, in centimetres, of a short rod, where $S \sim N(5, 0.25)$

L is the length, in centimetres, of a long rod, where $L \sim N(10, 1)$

Rods of either sort are joined to give longer lengths.

Find the probability that a length consisting of:

(i) 2 short rods and 4 long rods is greater than 52 cm

(ii) 3 short rods and 2 long rods is between 33 cm and 36 cm

(iii) 6 short rods is greater than a length consisting of 3 long rods.

14 Mr Smith has 5 dogs, 2 of which are male and 3 female. The masses of food they eat in any given week are normally distributed as follows:

	Mean (kg)	Standard deviation (kg)
Male	3.5	0.4
Female	2.5	0.3

Find the probability that the 2 males eat more than the 3 females in a particular week.

15 The time taken to carry out a standard service on a car of Type A may be modelled by a normal variable with mean 1 hour and standard deviation 10 minutes.

Assume that only one car is serviced at a time.

(i) Find the probability that it will take more than 6.5 hours to service 6 cars.

The time taken to carry out a standard service on a car of Type B is a normal variable with mean 1.5 hours and standard deviation 15 minutes.

(ii) Find the probability that 5 cars of Type B can be serviced more quickly than 8 cars of Type A.

16 The process of painting the bodywork of a mass-produced truck consists of giving it 1 coat of Paint A, 3 coats of Paint B and 2 coats of Paint C.

A record of the volumes of paint used for each truck has been kept for many years. It is found that, for each type of paint, the volume, in litres, of one coat is normally distributed with mean and standard deviation as shown in the table below.

	Mean (l)	Standard deviation (l)
One coat of Paint A	3.7	0.42
One coat of Paint B	1.3	0.15
One coat of Paint C	1.0	0.12

The volumes required of each type of paint are independent of each other.

(i) Find the mean and standard deviation of the total volume of paint used on a truck.

(ii) Find the percentage of trucks receiving:
    (a) less than 8.5 litres of paint
    (b) more than 10.0 litres of paint.

MULTIPLES OF NORMAL VARIABLES

Recall that, for the variable $aX + b$,

$$E(aX + b) = aE(X) + b, \operatorname{Var}(aX + b) = a^2\operatorname{Var}(X)$$

See page 26.

If X is **normally distributed** such that $X \sim N(\mu, \sigma^2)$, then the variable $aX + b$ is also normally distributed, where

$$aX + b \sim N(a\mu + b, a^2\sigma^2)$$

a and b are constants, where $a \neq 0$.

In particular, when $b = 0$,

$$aX \sim N(a\mu, a^2\sigma^2)$$

Recall that, for independent random variables X and Y,

$$E(aX + bY) = aE(X) + bE(Y), \operatorname{Var}(aX + bY) = a^2\operatorname{Var}(X) + b^2\operatorname{Var}(Y)$$

See page 28.

If X and Y are **independent normal variables** such that $X \sim N(\mu_1, \sigma_1^2)$ and $Y \sim N(\mu_2, \sigma_2^2)$, then

$$aX + bY \sim N(a\mu_1 + b\mu_2, a^2\sigma_1^2 + b^2\sigma_2^2)$$

$$aX - bY \sim N(a\mu_1 - b\mu_2, a^2\sigma_1^2 + b^2\sigma_2^2)$$

a and b are non-zero constants.

Example 2.22

The random variable X is normally distributed with mean 10 and standard deviation 2.

(i) State the distribution of $3X$.

(ii) State the distribution of $3X - 5$.

$X \sim N(10, 2^2)$

(i) $E(3X) = 3E(X) = 3 \times 10 = 30$ $a = 3$

   $\operatorname{Var}(3X) = 3^2\operatorname{Var}(X) = 9\operatorname{Var}(X) = 9 \times 2^2 = 36$

   So $3X \sim N(30, 36)$

(ii) $E(3X - 5) = 3E(X) - 5 = 3 \times 10 - 5 = 25$ $a = 3, b = -5$

   $\operatorname{Var}(3X - 5) = 3^2\operatorname{Var}(X) = 9\operatorname{Var}(X) = 9 \times 2^2 = 36$

   So $3X - 5 \sim N(25, 36)$

Example 2.23

A supermarket sells ice cream. During spring, the volume, in litres, of ice cream sold each day is normally distributed with mean 240 and standard deviation 20. During summer, there is a 30% increase in the daily sales of ice cream. Find the probability that the supermarket sells between 270 litres and 300 litres of ice cream on a summer's day.

Let X be the daily volume, in litres, of ice cream sold in spring, where $X \sim N(240, 20^2)$

Let Y be the daily volume, in litres, of ice cream sold in summer, where Y is 130% of X,

i.e. $Y = \dfrac{130}{100} X = 1.3X$

$E(Y) = 1.3E(X) = 1.3 \times 240 = 312$

$Var(Y) = 1.3^2 \, Var(X) = 1.69 \times 20^2 = 676$

So $Y \sim N(312, 676)$ with $\mu = 312$ and $\sigma = \sqrt{676} = 26$

$P(270 < Y < 300)$

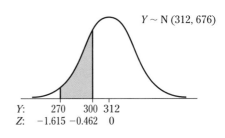

$$= P\left(\dfrac{270 - 312}{26} < Z < \dfrac{300 - 312}{26}\right)$$

$$= P(-1.615 < Z < -0.462)$$

$$= \Phi(1.615) - \Phi(0.462)$$

$$= 0.9468 - 0.6779$$

$$= 0.2689$$

$$= 0.269 \text{ (3 s.f.)}$$

Example 2.24

The weekly distance, in kilometres, driven by Mr Parry has a normal distribution with mean 512 and standard deviation 62. Independently, the weekly distance in kilometres driven by Mrs Parry has a normal distribution with mean 89 and standard deviation 7.4.

(i) Find the probability that, in a randomly chosen week, Mr Parry drives more than 5 times as far as Mrs Parry.

(ii) Find the mean and standard deviation of the total of the weekly distances in miles driven by Mr and Mrs Parry. Use the approximation 8 kilometres = 5 miles.

Cambridge Paper 7 Q4 J10(71)

Let X be the distance, in kilometres, driven by Mr Parry, where $X \sim N(512, 62^2)$

Let Y be the distance, in kilometres, travelled by Mrs Parry, where $Y \sim N(89, 7.4^2)$

(i) The variable '5 times Mrs Parry's distance' is denoted by $5Y$.

$E(5Y) = 5E(Y) = 5 \times 89 = 445$

$Var(5Y) = 5^2 Var(Y) = 25 \times 7.4^2 = 1369$

So $5Y \sim N(445, 1369)$

You require $P(X > 5Y)$, i.e. $P(X - 5Y > 0)$.

Let $D = X - 5Y$

You require $P(D > 0)$.

$E(D) = E(X) - E(5Y) = 512 - 445 = 67$

$Var(D) = Var(X) + Var(5Y) = 62^2 + 1369 = 5213$

$D \sim N(67, 5213)$ with $\mu = 67$ and $\sigma = \sqrt{5213}$

$$P(D > 0) = P\left(Z > \frac{0 - 67}{\sqrt{5213}}\right)$$

$$= P(Z > -0.928)$$

$$= \Phi(0.928)$$

$$= 0.8232$$

$$= 0.823 \text{ (3 s.f.)}$$

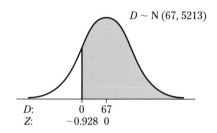

(ii) Let T be the total weekly distance, in kilometres, where $T = X + Y$

$E(T) = E(X) + E(Y) = 512 + 89 = 601$

$Var(T) = Var(X) + Var(Y) = 62^2 + 7.4^2 = 3898.76$

The total weekly distances are required in miles, so you need to convert km to miles.

Now $8 \text{ km} = 5 \text{ miles}$

so $1 \text{ km} = \frac{5}{8} \text{ mile}$

The total distance, in **miles**, is $\frac{5}{8} \times$ total distance in kilometres, i.e. $\frac{5}{8} T$.

$$E\left(\tfrac{5}{8} T\right) = \tfrac{5}{8} E(T) = \tfrac{5}{8} \times 601 = 375.625$$

So mean distance $= 376$ miles (3 s.f.)

$$Var\left(\tfrac{5}{8} T\right) = \left(\tfrac{5}{8}\right)^2 \times 3898.76 = 1522.95\ldots$$

So s.d. $= \sqrt{1522.95\ldots} = 39.02\ldots = 39.0$ miles (3 s.f.)

Example 2.25

The cost of electricity for a month in a certain town under scheme A consists of a fixed charge of 600 cents together with a charge of 5.52 cents per unit of electricity used. Stella uses scheme A. The number of units she uses in a month is normally distributed with mean 500 and variance 50.41.

(i) Find the mean and variance of the total cost of Stella's electricity in a randomly chosen month.

Under scheme B there is no fixed charge and the cost in cents for a month is normally distributed with mean 6600 and variance 421. Derek uses scheme B.

(ii) Find the probability that, in a randomly chosen month, Derek spends more than twice as much as Stella spends.

Cambridge Paper 7 Q4 N07

(i) Let X be the number of units of electricity used by Stella per month, where
$X \sim N(500, 50.41)$

Let T be the total cost, in cents, of Stella's electricity, where $T = 5.52X + 600$

Mean: $E(T) = 5.52E(X) + 600 = 5.52 \times 500 + 600 = 3360$

Variance: $\text{Var}(T) = 5.52^2\text{Var}(X) = 5.52^2 \times 50.41 = 1536.01\ldots = 1536$ (4 s.f.)

(ii) Let D be Derek's monthly cost, in cents, where $D \sim N(6600, 421)$

Also $T \sim N(3360, 1536)$

You require $P(D > 2T)$, i.e. $P(D - 2T > 0)$.

Let $Y = D - 2T$, so you require $P(Y > 0)$.

$$E(Y) = E(D) - 2E(T) = 6600 - 2 \times 3360 = -120$$

$$\text{Var}(Y) = \text{Var}(D) + 2^2\text{Var}(T) = 421 + 4 \times 1536 = 6565$$

$Y \sim N(-120, 6565)$ with $\mu = -120$ and $\sigma = \sqrt{6565}$

$P(Y > 0) = P\left(Z > \dfrac{0 - (-120)}{\sqrt{6565}}\right)$

$= P(Z > 1.481)$

$= 1 - \Phi(1.481)$

$= 1 - 0.9307$

$= 0.0693$

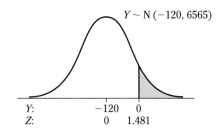

The probability that Derek spends more than twice as much as Stella spends is 0.0693

Distinguishing between sums and multiples of normal variables

You must take great care to distinguish between a **sum** of normal random variables and a **multiple** of a normal random variable.

For example, suppose X is the weight of a small loaf.

Then $X_1 + X_2 + X_3$ is the *total weight of three small loaves*. This is a **sum**.

If $X \sim N(\mu, \sigma^2)$, then $X_1 + X_2 + X_3 \sim N(3\mu, 3\sigma^2)$

Now consider a large economy loaf which is *three times the weight of a small loaf*; then the weight of an economy loaf is $3X$. This is a **multiple**.

If $X \sim N(\mu, \sigma^2)$, then $3X \sim N(3\mu, 9\sigma^2)$

In general, for $X \sim N(\mu, \sigma^2)$,

Sum: $X_1 + X_2 + \ldots + X_n \sim N(n\mu, n\sigma^2)$

Multiple: $nX \sim N(n\mu, n^2\sigma^2)$

Notice that the means are the same, but the variances are not. The distribution for the multiple has a greater spread.

Example 2.26

A soft-drinks manufacturer sells bottles of drinks in two sizes. The amounts, in millilitres, are normally distributed with mean and variance shown in the table.

	Mean (ml)	Variance (ml²)
Small bottle	252	4
Large bottle	1012	25

(i) A bottle of each size is selected at random. Find the probability that the large bottle contains less than 4 times the amount in a small bottle.

(ii) An inspector selects 1 large bottle and 4 small bottles at random. Find the probability that the amount in 1 large bottle is less than the total amount in the 4 small bottles.

Let S be the amount, in millilitres, in a small bottle, where $S \sim N(252, 4)$

Let L be the amount, in millilitres, in a large bottle, where $L \sim N(1012, 25)$

(i) To find the probability that the large bottle contains less than <u>four times the amount in a small bottle</u>, you need $P(L < 4S) = P(L - 4S < 0)$

$4S$ is a multiple.

Now $E(L - 4S) = E(L) - E(4S)$

$$= E(L) - 4E(S)$$

$$= 1012 - 4 \times 252$$

$$= 4$$

$\text{Var}(L - 4S) = \text{Var}(L) + \text{Var}(4S)$

$$= \text{Var}(L) + 4^2\text{Var}(S) \qquad \text{Remember + sign.}$$

$$= 25 + \mathbf{16} \times 4$$

$$= 89$$

So $L - 4S \sim N(4, 89)$ with $\mu = 4$ and $\sigma = \sqrt{89}$

$$P(L - 4S < 0) = P\left(Z < \frac{0 - 4}{\sqrt{89}}\right)$$

$$= P(Z < -0.424)$$

$$= 1 - \Phi(0.424)$$

$$= 1 - 0.6642$$

$$= 0.3358$$

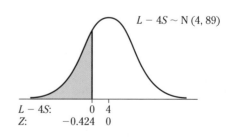

$L - 4S \sim N(4, 89)$

$L - 4S$:   0   4
Z: -0.424 0

The probability that a large bottle contains less than 4 times the amount in a small bottle is 0.336 (3 s.f.).

(ii) To find the probability that the large bottle contains less than <u>the total amount in 4 small bottles</u>, you need $P(L < S_1 + S_2 + S_3 + S_4)$

$$= P(L - (S_1 + S_2 + S_3 + S_4) < 0)$$

Now $E(L - (S_1 + S_2 + S_3 + S_4))$ $\longleftarrow$ $S_1 + S_2 + S_3 + S_4$ is a **sum**.

$\quad = E(L) - E(S_1 + S_2 + S_3 + S_4)$

$\quad = E(L) - 4E(S)$

$\quad = 1012 - 4 \times 252$

$\quad = 4$

$Var(L - (S_1 + S_2 + S_3 + S_4))$

$\quad = Var(L) + \mathbf{4}Var(S)$

$\qquad$ Remember $+$ sign.

$\quad = 25 + \mathbf{4} \times 4$

$\quad = 41$

So $L - (S_1 + S_2 + S_3 + S_4) \sim N(4, 41)$ with $\mu = 4$ and $\sigma = \sqrt{41}$

$P(L - (S_1 + S_2 + S_3 + S_4) < 0)$

$\quad = P\left(Z < \dfrac{0 - 4}{\sqrt{41}}\right)$

$\quad = P(Z < -0.625)$

$\quad = 1 - \Phi(0.625)$

$\quad = 1 - 0.7340$

$\quad = 0.2660$

$L - (S_1 + S_2 + S_3 + S_4)) \sim N(4, 41)$

$L - (S_1 + \cdots + S_4)$:   0   4

Z: -0.625 0

The probability that a large bottle contains less than the total amount in 4 small bottles is 0.266 (3 s.f.).

In the above example it was very important to distinguish between:

- the **multiple** of S in part (i)
- the **sum** of 4 independent observations of S in part (ii).

Note that

$$E(L - 4S) = 4$$
$$E(L - (S_1 + S_2 + S_3 + S_4)) = 4$$

The means are the same.

$$Var(L - 4S) = 89$$
$$Var(L - (S_1 + S_2 + S_3 + S_4)) = 41$$

The variances are different.

Example 2.27

Climbing ropes produced by a manufacturer have breaking strengths which are normally distributed with mean 160 kg and standard deviation 11.3 kg. A group of climbers have weights which are normally distributed with mean 66.3 kg and standard deviation 7.1 kg.

(i) Find the probability that a rope chosen randomly will break under the combined weight of 2 climbers chosen randomly.

Each climber carries, in a rucksack, equipment amounting to half his own weight.

(ii) Find the mean and the variance of the combined weight of a climber and his rucksack.

(iii) Find the probability that the combined weight of a climber and his rucksack is greater than 87 kg.

Cambridge Paper 7 Q5 N06

Let R be the breaking strength, in kilograms, of a rope, where $R \sim N(160, 11.3^2)$

Let W be the weight, in kilograms, of a climber, where $W \sim N(66.3, 7.1^2)$

(i) The rope will break if the weight of the 2 men is greater than the rope's breaking strength, i.e. if $W_1 + W_2 > R$

So you require $P(W_1 + W_2 - R > 0)$.

$W_1 + W_2$ is a **sum**. Do <u>not</u> consider the multiple $2W$.

Let $X = W_1 + W_2 - R$

$$E(X) = E(W_1) + E(W_2) - E(R)$$

$$= 66.3 + 66.3 - 160$$

$$= -27.4$$

$$Var(X) = Var(W_1) + Var(W_2) + Var(R)$$

$$= 7.1^2 + 7.1^2 + 11.3^2$$

$$= 228.51$$

$X \sim N(-27.4, 228.51)$ $\mu = -27.4, \sigma = \sqrt{228.51}$

$$P(X > 0) = P\left(Z > \frac{0 - (-27.4)}{\sqrt{228.51}}\right)$$

$$= P(Z > 1.813)$$

$$= 1 - \Phi(1.813)$$

$$= 1 - 0.9651$$

$$= 0.0349$$

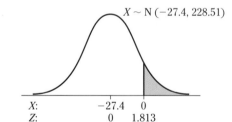

The probability that the rope will break is 0.0349

Note: You could have found $P(R - (W_1 + W_2) < 0)$.

(ii) Let C be the combined weight of a climber and his sack, where $C = 1.5W$

This is a **multiple** of the variable W.

Mean: $E(C) = 1.5E(W) = 1.5 \times 66.3 = 99.45$

Variance: $Var(C) = 1.5^2Var(W) = 2.25 \times 7.1^2 = 113.4225$

(iii) $C \sim N(99.45, 113.4225)$ $\mu = 99.45, \sigma = \sqrt{113.4225}$

$$P(C > 87) = P\left(Z > \frac{87 - 99.45}{\sqrt{113.4225}}\right)$$

$$= P(Z > -1.169)$$

$$= \Phi(1.169)$$

$$= 0.8788$$

$$= 0.879 \text{ (3 s.f.)}$$

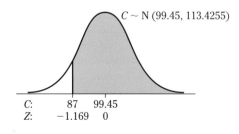

Exercise 2d

1 The random variable X is normally distributed with mean 15 and standard deviation 3.5

State the distribution of:

(i) $4X$ (ii) $\frac{1}{3}X - 5$ (iii) $20 - 2X$.

2 X and Y are independent normal variables such that $X \sim N(40, 12)$ and $Y \sim N(60, 15)$

Find:

(i) $P(2X + Y > 130)$ (ii) $P(3X - 2Y < 20)$.

3 X and Y are independent normal variables such that $X \sim N(100, 100)$ and $Y \sim N(65, 64)$

Find:

(i) $P(X < 2Y)$ (ii) $P\left(Y > \frac{1}{2}X\right)$.

4 X and Y are random variables, where $X \sim N(100, 8)$ and $Y \sim N(55, 10)$

Find the probability that a random observation from the population of X is more than twice the value of a random observation from the population of Y.

5 The time, S minutes, taken by Simon to do his mathematics homework is normally distributed with mean 50 and standard deviation 10. The time, B minutes, taken by Belinda to do her mathematics homework is normally distributed with mean 30 and standard deviation 5.

The variables S and B are independent.

(i) Find the probability that, for a particular homework, Simon takes more than twice as long as Belinda.

(ii) Find the probability that the total time Belinda takes to do her mathematics homework on Monday and Thursday is less than the time Simon takes to do his mathematics homework on Monday.

6 The thickness, X centimetres, of a hardback book may be modelled by a normal distribution with mean 4.9 and variance 1.920.

The thickness, Y centimetres, of a paperback book may be modelled by a normal distribution with mean 2.0 and variance 0.730.

Find the probability that a randomly chosen paperback book is less than half as thick as a randomly chosen hardback book.

7 The random variable X is distributed $N(60, 8^2)$. The random variable Y is distributed $N(52, 6^2)$.

Find the probability that a random observation from X is more than 25% greater than a random observation from Y.

8 A certain school holds cake sales to raise money for charity. It is found that the amount of money raised by a cake sale at the school may be modelled by a normal distribution with mean \$23.50 and standard deviation \$3.40.

Class 4T wants to raise \$40 for some sports equipment.

The Parents' Association has offered to match the money raised in 4T's cake sale. For example if the cake sale raises \$20, the Parents' Association will contribute \$20.

However, a student in the class says that more money will be raised if 4T holds two cake sales instead of accepting this offer.

(i) Find the probability of raising at least \$40 by each of these methods.

(ii) Which option would you advise the class to choose? Give a reason for your answer.

Summary

Linear functions of a variable

For any random variable X and constants a and b,

$$E(aX) = aE(X)$$
$$E(aX + b) = aE(X) + b$$
$$Var(aX) = a^2Var(X)$$
$$Var(aX + b) = a^2Var(X)$$

Linear combinations of variable

For two random variables X and Y,

$$\mathrm{E}(aX + bY) = a\mathrm{E}(X) + b\mathrm{E}(Y)$$

$$\mathrm{E}(aX - bY) = a\mathrm{E}(X) - b\mathrm{E}(Y)$$

If X and Y are independent,

$$\mathrm{Var}(aX + bY) = a^2\mathrm{Var}(X) + b^2\mathrm{Var}(Y)$$

$$\mathrm{Var}(aX - bY) = a^2\mathrm{Var}(X) + b^2\mathrm{Var}(Y)$$

For n observations of the random variable X,

$$\mathrm{E}(X_1 + X_2 + \ldots + X_n) = n\mathrm{E}(X)$$

If the observations are independent,

$$\mathrm{Var}(X_1 + X_2 + \ldots + X_n) = n\mathrm{Var}(X)$$

Combinations of Poisson variables

For independent Poisson variables X and Y, where $X \sim \mathrm{Po}(\lambda_1)$ and $Y \sim \mathrm{Po}(\lambda_2)$,

$$X + Y \sim \mathrm{Po}(\lambda_1 + \lambda_2)$$

Combinations of normal variables

For independent normal variables $X \sim \mathrm{N}(\mu_1, \sigma_1^2)$ and $Y \sim \mathrm{N}(\mu_2, \sigma_2^2)$ and constants a and b:

Sums

$$X + Y \sim \mathrm{N}(\mu_1 + \mu_2, \sigma_1^2 + \sigma_2^2)$$

$$X_1 + X_2 + \ldots + X_n \sim \mathrm{N}(\mu_1 + \mu_2 + \ldots + \mu_n, \sigma_1^2 + \sigma_2^2 + \ldots + \sigma_n^2)$$

For n observations from the **same** normal variable X

$$X_1 + X_2 + \ldots + X_n \sim \mathrm{N}(n\mu, n\sigma^2)$$

Differences

$$X - Y \sim \mathrm{N}(\mu_1 - \mu_2, \sigma_1^2 + \sigma_2^2)$$

Multiples

$$aX \sim \mathrm{N}(a\mu, a^2\sigma^2)$$

$$aX + b \sim \mathrm{N}(a\mu + b, a^2\sigma^2)$$

$$aX + bY \sim \mathrm{N}(a\mu_1 + b\mu_2, a^2\sigma_1^2 + b^2\sigma_2^2)$$

$$aX - bY \sim \mathrm{N}(a\mu_1 - b\mu_2, a^2\sigma_1^2 + b^2\sigma_2^2)$$

Mixed Exercise 2

1 X and Y are independent random variables. The random variable X has mean 27 and variance 25. The random variable Y has mean 62 and variance 36.

Find the mean and standard deviation of the following variables:

(i) $A = 3X + 2$

(ii) $B = X + 2Y$

(iii) $C = 5X - Y + 10$

(iv) $D = X_1 + X_2 + X_3 + Y_1 + Y_2$, where X_1, X_2 and X_3 are independent observation of X, and Y_1 and Y_2 are independent observations of Y.

2 A mathematics module is assessed by an examination and by coursework. The examination makes up 75% of the total assessment and the coursework makes up 25%. Examination marks, X, are distributed with mean 53.2 and standard deviation 9.3. Coursework marks, Y, are distributed with mean 78.0 and standard deviation 5.1. Examination marks and coursework marks are independent. Find the mean and standard deviation of the combined mark $0.75X + 0.25Y$.

Cambridge Paper 7 Q2 J06

3 X and Y are independent random variables such that $X \sim Po(2.5)$ and $Y \sim Po(1.8)$

Find:

(i) $P(X + Y = 7)$ (ii) $P(X + Y \leqslant 3)$.

4 Computer breakdowns occur randomly on average once every 48 hours of use.

(i) Calculate the probability that there will be fewer than 4 breakdowns in 60 hours of use.

(ii) Find the probability that the number of breakdowns in one year (8760 hours) of use is more than 200.

(iii) Independently of the computer breaking down, the computer operator receives phone calls on average twice in every 24-hour period. Find the probability that the total number of phone calls and computer breakdowns in a 60-hour period is exactly 4.

Cambridge Paper 7 Q6 J03

5 In their football matches, Rovers score goals independently and at random times. Their average rate of scoring is 2.3 goals per match.

(i) State the expected number of goals that Rovers will score in the first half of a match.

(ii) Find the probability that Rovers will not score any goals in the first half of a match but will score one or more goals in the second half of the match.

(iii) Football matches last for 90 minutes. In a particular match, Rovers score one goal in the first 30 minutes. Find the probability that they will score at least one further goal in the remaining 60 minutes.

Independently of the number of goals scored by Rovers, the number of goals scored per football match by United has a Poisson distribution with mean 1.8.

(iv) Find the probability that a total of at least 3 goals will be scored in a particular match when Rovers play United.

Cambridge Paper 7 Q6 N08

6 At a petrol station cars arrive independently and at random times at constant average rates of 8 cars per hour travelling east and 5 cars per hour travelling west.

(i) Find the probability that, in a quarter-hour period,

(a) one or more cars travelling east and one or more cars travelling west will arrive,

(b) a total of 2 or more cars will arrive.

(ii) Find the approximate probability that, in a 12-hour period, a total of more than 175 cars will arrive.

Cambridge Paper 7 Q6 J05

7 The mass of a certain grade of apple is normally distributed with mean 120 g and standard deviation 10 g.

(i) An apple of this grade is selected at random. Find the probability that its mass lies between 100.5 g and 124 g.

(ii) Four apples of this grade are selected at random. Find the probability that their total mass exceeds 505 g.

8 X and Y are independent random variables such that $X \sim N(130, 36)$ and $Y \sim N(95, 15)$

The random variable W is defined by $W = X_1 + X_2 + 2Y$, where X_1 and X_2 are two independent observations of X.

(i) Find the mean and variance of W.

(ii) Find $P(444 < W < 468)$.

9 Tien throws a ball. The distance it travels can be modelled by a normal distribution with mean 20 m and variance 9 m^2. His younger sister Su Chen also throws a ball and the distance her ball travels can be modelled by a normal distribution with mean 14 m and variance 12 m^2. Su Chen is allowed to add 5 metres on to her distance and call it her 'upgraded distance'. Find the probability that Tien's distance is larger than Su Chen's upgraded distance.

Cambridge Paper 7 Q3 N03

10 Bottles of wine are stacked in racks of 12. The weights of these bottles are normally distributed with mean 1.3 kg and standard deviation 0.06 kg. The weights of the empty racks are normally distributed with mean 2 kg and standard deviation 0.3 kg.

(i) Find the probability that the total weight of a full rack of 12 bottles of wine is between 17 kg and 18 kg.

(ii) Two bottles of wine are chosen at random. Find the probability that they differ in weight by more than 0.05 kg.

Cambridge Paper 7 Q7 N02

11 The volume of liquid in cans of cola is normally distributed with mean 330 millilitres and standard deviation 5.2 millilitres. The volume of liquid in bottles of tonic water is normally distributed with mean 500 millilitres and standard deviation 7.1 millilitres.

(i) Find the probability that 3 randomly chosen cans of cola contain less liquid than 2 randomly chosen bottles of tonic water.

(ii) A new drink is made by mixing the contents of 2 cans of cola with half a bottle of tonic water. Find the probability that the volume of the new drink is more than 900 millilitres.

Cambridge Paper 7 Q7 N09(71)

12 A journey in a certain car consists of two stages with a stop for filling up with fuel after the first stage. The length of time, T minutes, taken for each stage has a normal distribution with mean 74 and standard deviation 7.3. The length of time, F minutes, it takes to fill up with fuel has a normal distribution with mean 5 and standard deviation 1.7. The length of time it takes to pay for the fuel is exactly 4 minutes. The variables T and F are independent and the times for the two stages are independent of each other.

(i) Find the probability that the total time for the journey is less than 154 minutes.

(ii) A second car has a fuel tank with exactly twice the capacity of the first car. Find the mean and variance of this car's fuel fill-up time.

(iii) This second car's time for each stage of the journey follows a normal distribution with mean 69 minutes and standard deviation 5.2 minutes. The length of time it takes to pay for the fuel for this car is also exactly 4 minutes. Find the probability that the total time for the journey taken by the first car is more than the total time taken by the second car.

Cambridge Paper 7 Q7 N05

3 Continuous probability distributions

In this chapter you will learn

- about the probability density function of a continuous random variable X
- how to find a probability by calculating the area under the graph of $y = f(x)$
- how to find the median and quartiles of X
- how to find the mean and standard deviation of X

CONTINUOUS RANDOM VARIABLES

You have already seen in S1 chapter 4 that a **discrete random variable** is a variable that can take individual values each with a given probability. For example, the probability distribution of X, the score on a fair cubical die is as follows:

x	1	2	3	4	5	6
$P(X = x)$	$\frac{1}{6}$	$\frac{1}{6}$	$\frac{1}{6}$	$\frac{1}{6}$	$\frac{1}{6}$	$\frac{1}{6}$

So, for instance, $P(X = 3) = \frac{1}{6}$.

By contrast, a **continuous random variable** cannot take precise values but can be defined only within a specified interval. For example, when a boy's height is given as 126 cm, measured to the nearest cm, this means that the height could be anywhere in the interval 125.5 cm $\leqslant$ height $<$ 126.5 cm. Continuous variables are associated with measurements of characteristics such as time, mass or length.

Probability density function

A continuous random variable X is defined by its **probability density function** f, and it can be illustrated by the graph of $y = f(x)$, for example:

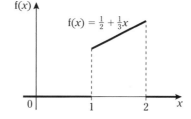

$$f(x) = \begin{cases} \frac{1}{2} + \frac{1}{3}x & 1 \leqslant x \leqslant 2 \\ 0 & \text{otherwise} \end{cases}$$

Note that a thicker line is drawn on the x-axis outside the interval $1 \leqslant x \leqslant 2$ to show that $y = 0$ for these values of x.

Finding probabilities

If a continuous random variable X has probability density function f then the **probability** that x lies in the interval $a \leqslant x \leqslant b$ is given by the **area** under the graph of $y = f(x)$ between $x = a$ and $x = b$.

This area can be found by **integration**, or sometimes by geometry.

Note:

- Since probabilities cannot be negative, the graph of $y = f(x)$ never goes below the x-axis.
- Since the total probability is 1, the total area under the graph is 1.

For a **continuous random variable** X with **probability density function** given by $f(x)$ for $a \leqslant x \leqslant b$,

$$P(x_1 \leqslant X \leqslant x_2) = \int_{x_1}^{x_2} f(x)\,dx$$

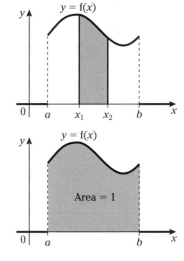

Since the total area under the curve is 1,

$$\int_{\text{all } x} f(x)\,dx = \int_a^b f(x)\,dx = 1$$

Note: This compares with discrete random variables, where $\sum_{i=1}^n p_i = 1$ (S1 Chapter 4)

Note that this is different from calculating probabilities for discrete variables (such as variables having a binomial or a Poisson distribution).

For discrete random variables, $P(X = 3)$, for example, has a definite value which is usually not zero. This is different for continuous random variables such as time, for example, where a measurement of 3 seconds correct to the nearest tenth of a second could be anywhere in the interval 2.95 seconds $\leqslant X <$ 3.05 seconds. This interval becomes narrower and narrower as you try to approach the instant of time of 3 seconds and the probability that X takes the *exact* value of 3 is zero.

In fact, for any continuous random variable X

$P(X = k) = 0$, where k is any constant.

You will recall that for discrete random variables, $P(X < 3)$ is not the same as $P(X \leqslant 3)$. However, for continuous variables it is not possible to distinguish between $P(X < 3)$ and $P(X \leqslant 3)$, nor between $P(X > 3)$ and $P(X \geqslant 3)$.

In general, if X is a continuous random variable it is not possible to distinguish between the following probabilities, but if X is discrete they would be different:

$$P(x_1 < X < x_2), \qquad P(x_1 \leqslant X < x_2), \qquad P(x_1 < X \leqslant x_2), \qquad P(x_1 \leqslant X \leqslant x_2)$$

This difference between discrete and continuous random variables explains the need for a continuity correction when using the normal distribution as an approximation to the binomial distribution (S1 chapter 6) or Poisson distribution (S2 chapter 1).

Integration note

In the examination you may be required to integrate functions described in the integration sections of P1 and P3. However, so that you can study this chapter before all the integration methods have been covered, P3 functions occur only in the miscellaneous examples and Mixed Exercise 3 at the end of the chapter.

Example 3.1

The continuous random variable X has probability density function given by

$$f(x) = \begin{cases} kx^2 & 1 \leqslant x \leqslant 4 \\ 0 & \text{otherwise} \end{cases}$$

where k is a constant.

(i) Draw a sketch of the probability density function.

(ii) Find the value of k. (iii) Calculate $P(2.5 \leqslant X \leqslant 3.5)$.

(i) The graph of $y = f(x)$ is part of a quadratic curve.
When $x = 1$, $y = k$; when $x = 4$, $y = 16k$.

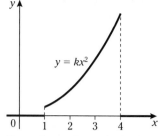

(ii) Total area under curve

$$= \int_{\text{all } x} f(x)\, dx$$

$$= \int_1^4 kx^2\, dx$$

$$= k \left[\frac{x^3}{3} \right]_1^4$$

It is easier to substitute the limits if you take out numerical factors first.

$$= \frac{k}{3} \left[x^3 \right]_1^4$$

$$= \frac{k}{3} (4^3 - 1^3)$$

$$= 21k$$

Total area $= 1$

so $\qquad 21k = 1$

$$k = \frac{1}{21}$$

Remember to draw a thicker line on the x-axis to show that $y = 0$ for all x-values outside the interval $1 \leqslant x \leqslant 4$.

(iii) $f(x) = \frac{1}{21} x^2$

$$P(2.5 \leqslant X \leqslant 3.5) = \int_{2.5}^{3.5} \frac{1}{21} x^2\, dx$$

$$= \frac{1}{21} \left[\frac{x^3}{3} \right]_{2.5}^{3.5}$$

$$= \frac{1}{63} \left[x^3 \right]_{2.5}^{3.5}$$

$$= \frac{1}{63} (3.5^3 - 2.5^3)$$

$$= 0.4325\ldots$$

$$= 0.433 \ (3 \text{ s.f.})$$

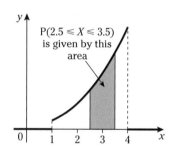

$P(2.5 \leqslant X \leqslant 3.5)$ is given by this area

Example 3.2

The random variable X denotes the mass, in kilograms, of a substance produced in an industrial process. The probability density function of X is given by

$$f(x) = \begin{cases} \frac{1}{36} x(6 - x) & 0 \leqslant x \leqslant 6 \\ 0 & \text{otherwise} \end{cases}$$

Calculate the probability that more than 5 kg of the substance is produced in the industrial process.

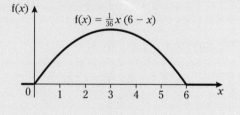

$f(x) = \frac{1}{36} x(6 - x)$

$$P(X > 5) = \int_5^6 \frac{1}{36} x(6 - x)\, dx$$

$$= \frac{1}{36} \int_5^6 (6x - x^2)\, dx$$

$$= \frac{1}{36} \left[3x^2 - \frac{x^3}{3} \right]_5^6$$

$$= \frac{1}{36} \left(\left(3 \times 6^2 - \frac{6^3}{3} \right) - \left(3 \times 5^2 - \frac{5^3}{3} \right) \right)$$

$$= \frac{2}{27}$$

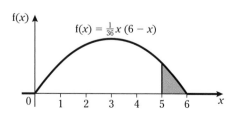

$f(x) = \frac{1}{36} x(6 - x)$

If you give your final answer in decimals, round to 3 significant figures.

Note: The following alternative method may sometimes be easier.

$$P(X > 5) = 1 - P(X \leqslant 5)$$
$$= 1 - \frac{1}{36}\left[3x^2 - \frac{x^3}{3}\right]_0^5$$
$$= 1 - \frac{1}{36}\left(\left(3 \times 5^2 - \frac{5^3}{3}\right) - 0\right)$$
$$= 1 - \frac{25}{27}$$
$$= \frac{2}{27}$$

Example 3.3

The error, in grams, made by a set of weighing scales may be modelled by the random variable X with probability density function given by

$$f(x) = \begin{cases} k & -3 \leqslant x \leqslant 7 \\ 0 & \text{otherwise} \end{cases}$$

where k is a constant.

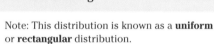

(i) Find the value of k.

(ii) Find the probability that an error is positive.

(iii) Given that an error is positive, find the probability that the error is less than $4\,\text{g}$.

(iv) Find the probability that the **magnitude** of an error is less than $2\,\text{g}$.

(i) Total area under graph $= 10k$

so $10k = 1$

$k = 0.1$

(ii) $f(x) = 0.1$

P(error is positive)

$= P(X > 0)$

$= 0.1 \times 7$

$= 0.7$

(iii) $P(X < 4 | X > 0)$

$= \dfrac{P(0 < X < 4)}{P(X > 0)}$

$= \dfrac{0.4}{0.7}$

$= \dfrac{4}{7}$

(iv) P(magnitude of error is more than $2\,\text{g}$)

$= P(|X| > 2)$

$= P(X < -2) + P(X > 2)$

$= 0.1 \times 1 + 0.1 \times 5$

$= 0.6$

Note: This distribution is known as a **uniform** or **rectangular** distribution.

There is no need to integrate as you can find the area easily using geometry.

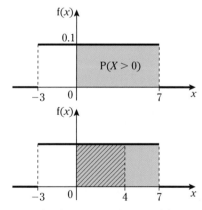

The magnitude of an error is the size of the error, irrespective of whether it is positive or negative. For example, if the error is -2.5, the magnitude is 2.5.

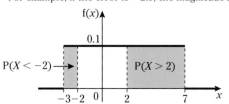

Example 3.4

The continuous random variable T has probability density function given by

$$f(t) = \begin{cases} \dfrac{k}{t^4} & t \geq 1 \\ 0 & \text{otherwise} \end{cases}$$

where k is a constant.

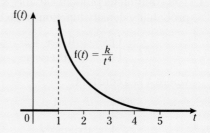

(i) Find the value of k.

(ii) Calculate $P(T < 1.5)$.

(i) Total area under curve

$$= \int_{\text{all } t} f(t)\, dt$$

$$= \int_1^\infty \frac{k}{t^4}\, dt \qquad \text{The upper limit is infinity.}$$

$$= k \int_1^\infty t^{-4}\, dt$$

$$= k \left[\frac{t^{-3}}{-3} \right]_1^\infty$$

$$= -\frac{k}{3} \left[\frac{1}{t^3} \right]_1^\infty$$

$$= -\frac{k}{3}(0 - 1) \qquad \frac{1}{t^3} \to 0 \text{ when } t \to \infty$$

$$= \frac{k}{3}$$

Total area $= 1$

so $\quad \dfrac{k}{3} = 1$

$\quad k = 3$

(ii) $\quad f(t) = \dfrac{3}{t^4} = 3t^{-4}$

$$P(T < 1.5) = 3 \int_1^{1.5} t^{-4}\, dt$$

$$= 3 \left[\frac{t^{-3}}{-3} \right]_1^{1.5}$$

$$= -\left[\frac{1}{t^3} \right]_1^{1.5}$$

$$= -\left(\frac{1}{1.5^3} - 1 \right)$$

$$= 0.7037\ldots$$

$$= 0.704 \text{ (3 s.f.)}$$

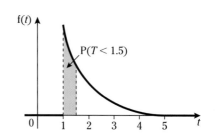

Exercise 3a

1 The continuous random variable X has probability density function given by

$$f(x) = \begin{cases} kx^2 & 0 \le x \le 2 \\ 0 & \text{otherwise} \end{cases}$$

where k is a constant.

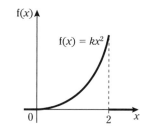

(i) Find the value of k.

(ii) Find $P(X \ge 1)$.

(iii) Find $P(0.5 \le X \le 1.5)$.

2 The continuous random variable X has probability density function given by

$$f(x) = \begin{cases} k & -2 \le x \le 3 \\ 0 & \text{otherwise} \end{cases}$$

where k is a constant.

(i) Sketch the probability density function of X.

(ii) Find the value of k.

(iii) Find $P(-1.6 \le X \le 2.1)$.

(iv) Find $P(-2.5 < X < 2.5)$.

Note: $P(-2.5 < X < 2.5)$ can be written $P(|X| < 2.5)$.

(v) (a) Find $P(X > 1)$.

(b) Given than X is greater than 1, find the probability that X is less than 1.5

3 The continuous random variable X has probability density function given by

$$f(x) = \begin{cases} k(4 - x) & 1 \le x \le 3 \\ 0 & \text{otherwise} \end{cases}$$

where k is a constant.

(i) Find the value of k.

(ii) Sketch the probability density function of X.

(iii) Find $P(X > 2)$.

(iv) Find $P(1.2 \le X \le 2.4)$.

4 The continuous random variable X has probability density function given by

$$f(x) = \begin{cases} k(x + 2)^2 & 0 \le x \le 2 \\ 0 & \text{otherwise} \end{cases}$$

where k is a constant.

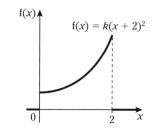

(i) Find the value of k.

(ii) Find $P(0 \le X \le 1)$ and hence find $P(X > 1)$.

5 The continuous random variable X has probability density function given by

$$f(x) = \begin{cases} kx^3 & 0 \le x \le c \\ 0 & \text{otherwise} \end{cases}$$

where k and c are constants.

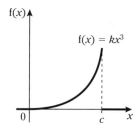

Given that $P\left(X \le \frac{1}{2}\right) = \frac{1}{16}$, find c and k.

6 The continuous random variable X has probability density function given by

$$f(x) = \begin{cases} kx & 0 \le x \le 4 \\ 0 & \text{otherwise} \end{cases}$$

where k is a constant.

(i) Find the value of k.

(ii) Sketch the probability density function of X.

(iii) Find $P(1 \le X \le 2.5)$.

7 The delay, in hours, of a flight from Chicago can be modelled by the continuous random variable X with probability density function given by

$$f(x) = \begin{cases} \frac{1}{50}(10 - x) & 0 \le x \le 10 \\ 0 & \text{otherwise} \end{cases}$$

(i) Find the probability that the delay will be less than 4 h.

(ii) Find the probability that the delay will be between 2 h and 6 h.

8 The continuous random variable X has probability density function given by

$$f(x) = \begin{cases} k & a \leqslant x \leqslant 10 \\ 0 & \text{otherwise} \end{cases}$$

where k is a constant.

It is known that $P(5 < X < 8) = 0.5$
Find the values of k and a.

9 The continuous random variable X has probability density function given by

$$f(x) = \begin{cases} k\sqrt{x} & 1 \leqslant x \leqslant 9 \\ 0 & \text{otherwise} \end{cases}$$

where k is a constant.

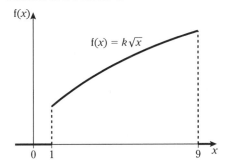

(i) Find the value of k.

(ii) Find $P(4 \leqslant X \leqslant 9)$.

10 The continuous random variable X has probability density function given by

$$f(x) = \begin{cases} \dfrac{k}{x^2} & 1 \leqslant x \leqslant 3 \\ 0 & \text{otherwise} \end{cases}$$

where k is a constant.

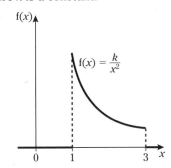

(i) Find the value of k.

(ii) Find $P(X \leqslant 2)$.

11 The continuous random variable X has probability density function given by

$$f(x) = \begin{cases} \dfrac{2}{x^3} & x \geqslant 1 \\ 0 & \text{otherwise} \end{cases}$$

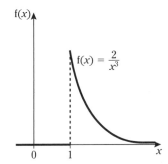

Find $P(X > 2)$.

12 Explain, with a reason, whether each of these functions could be the probability density function of a random variable X.

(i)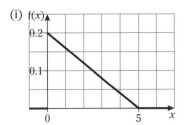

(ii)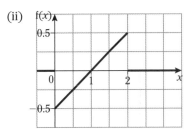

(iii)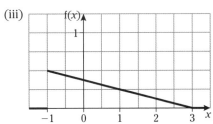

13 The random variable X has probability density function given by

$$f(x) = \begin{cases} k(x^2 - 4) & 0 \leqslant x \leqslant 2 \\ 0 & \text{otherwise} \end{cases}$$

where k is a constant.

Find the value of k.

14 The random variable X has probability density function given by

$$f(x) = \begin{cases} \dfrac{k}{x^4} & x \geqslant 2 \\ 0 & \text{otherwise} \end{cases}$$

where k is a constant.

(i) Find the value of k.

(ii) Find the probability that X is greater than 3.

15 The random variable X has probability density function given by

$$f(x) = \begin{cases} a(b - x) & 0 \leqslant x \leqslant b \\ 0 & \text{otherwise} \end{cases}$$

where a and b are constants.

(i) Show that $a = \dfrac{2}{b^2}$

(ii) Given that $a = \frac{1}{8}$, find $P(2 \leqslant X \leqslant 3)$.

MEDIAN AND QUARTILES

Consider the continuous variable X, defined by its probability density function f for $a \leqslant x \leqslant b$.

Median

The **median** of the distribution of X is defined as the value which divides the area under $y = f(x)$ in half.

Denoting the **median** by m,

$$P(X \leqslant m) = \int_a^m f(x)\,dx = 0.5$$

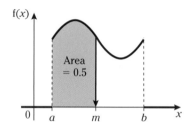

Quartiles

The quartiles of the distribution of X are the values which, together with the median, divide the area under $y = f(x)$ into quarters.

Denoting the **lower quartile** by q_1,

$$P(X \leqslant q_1) = \int_a^{q_1} f(x)\,dx = 0.25$$

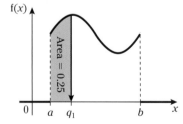

Denoting the **upper quartile** by q_3,

$$P(X \leqslant q_3) = \int_a^{q_3} f(x)\,dx = 0.75$$

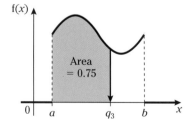

Remember that

interquartile range = upper quartile − lower quartile

$$= q_3 - q_1$$

Example 3.5

The continuous random variable X has probability density function given by

$$f(x) = \begin{cases} \frac{1}{8}x & 0 \leqslant x \leqslant 4 \\ 0 & \text{otherwise} \end{cases}$$

(i) Find the median of X.   (ii) Find the interquartile range.

(i) Denoting the median by m,

$$P(X \leqslant m) = \int_{\text{all }x} f(x)\,dx$$

$$= \int_0^m \frac{1}{8}x\,dx$$

$$= \frac{1}{8}\left[\frac{x^2}{2}\right]_0^m$$

$$= \frac{1}{16}\left[x^2\right]_0^m$$

$$= \frac{1}{16}(m^2 - 0)$$

$$= \frac{1}{16}m^2$$

Since $P(X \leqslant m) = 0.5$

$$\frac{1}{16}m^2 = 0.5$$

$$m^2 = 8$$

$$m = \sqrt{8}$$

Take the positive square root since m lies between 0 and 4.

$$= 2.828\ldots$$

$$= 2.83 \text{ (3 s.f.)}$$

(ii) For the lower quartile q_1,

$$P(X \leqslant q_1) = \int_0^{q_1} \frac{1}{8}x\,dx$$

The integration is the same as in part (i), with m replaced by q_1.

$$= \frac{1}{16}q_1{}^2$$

Since $P(X \leqslant q_1) = 0.25$,

$$\frac{1}{16}q_1{}^2 = 0.25$$

$$q_1{}^2 = 4$$

$$q_1 = 2$$

Similarly, since $P(X \leqslant q_3) = 0.75$,

The integration is the same as in part (i), with m replaced by q_3.

$$\frac{1}{16}q_3{}^2 = 0.75$$

$$q_3{}^2 = 12$$

$$q_3 = \sqrt{12}$$

Interquartile range $= q_3 - q_1$

$$= \sqrt{12} - 2$$

$$= 1.464\ldots = 1.46 \text{ (3 s.f.)}$$

Note:

In Example 3.5, since $y = f(x)$ is a straight line, the median and quartiles could be found using geometry.

For example, to find the median:

When $x = m$, $y = \frac{1}{8}m$

$$P(X \leqslant m) = \text{area of shaded triangle}$$

Area = 0.5

$$= \frac{1}{2} \times m \times \frac{1}{8}m$$

$$= \frac{1}{16}m^2$$

Since $P(X \leqslant m) = 0.5$

$$\frac{1}{16}m^2 = 0.5$$

$$m^2 = 8$$

$$m = \sqrt{8} \text{ (as in part (i) on the previous page).}$$

Example 3.6

The continuous random variable X has probability density function given by

$$f(x) = \begin{cases} \dfrac{8}{3x^3} & 1 \leqslant x \leqslant 2 \\ 0 & \text{otherwise} \end{cases}$$

Find the median of X.

Denoting the median by m,

$$P(X \leqslant m) = \int_1^m f(x)\, dx$$

$$= \int_1^m \frac{8}{3x^3}\, dx$$

$$= \frac{8}{3} \int_1^m x^{-3}\, dx \qquad \text{Write } \frac{1}{x^3} \text{ as } x^{-3}.$$

$$= \frac{8}{3} \left[\frac{x^{-2}}{-2} \right]_1^m \qquad \text{Take care with the negatives when integrating.}$$

$$= -\frac{4}{3} \left[\frac{1}{x^2} \right]_1^m$$

$$= -\frac{4}{3} \left(\frac{1}{m^2} - 1 \right)$$

$$= \frac{4}{3} \left(1 - \frac{1}{m^2} \right)$$

Since $P(X \leqslant m) = 0.5$

$$\frac{4}{3} \left(1 - \frac{1}{m^2} \right) = 0.5$$

$$1 - \frac{1}{m^2} = 0.375$$

$$\frac{1}{m^2} = 0.625$$

$$m^2 = \frac{1}{0.625} = 1.6$$

$$m = \sqrt{1.6} = 1.264\ldots = 1.26 \text{ (3 s.f.)} \qquad \text{Reject } m = -\sqrt{1.6}, \text{ since median lies between 1 and 2}$$

Example 3.7

The continuous random variable X has probability density function given by

$$f(x) = \begin{cases} \frac{1}{8}(4-x) & 0 \leqslant x \leqslant 4 \\ 0 & \text{otherwise} \end{cases}$$

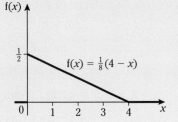

(i) Find $P(X > 2)$ and deduce the value of the upper quartile.

(ii) Find the median of X.

(i) *$P(X > 2)$ is given by the area of the triangle (shown shaded in the diagram).*

When $x = 2$, $y = \frac{1}{8}(4-2) = \frac{1}{4}$

Using geometry, area of triangle $= \frac{1}{2} \times 2 \times \frac{1}{4} = \frac{1}{4}$

So $P(X > 2) = 0.25$

$P(X \leqslant 2) = 0.75$, so 2 is the upper quartile of X.

Note: This uses the fact that the area **above** the upper quartile is 0.25.

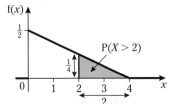

(ii) $P(X \leqslant m) = \displaystyle\int_0^m f(x)\,dx$

$$= \int_0^m \frac{1}{8}(4-x)\,dx$$

$$= \frac{1}{8}\int_0^m (4-x)\,dx$$

$$= \frac{1}{8}\left[4x - \frac{x^2}{2}\right]_0^m$$

$$= \frac{1}{8}\left(4m - \frac{m^2}{2} - 0\right)$$

$$= \frac{1}{8}\left(4m - \frac{m^2}{2}\right)$$

Since $P(X \leqslant m) = 0.5$,

$$\frac{1}{8}\left(4m - \frac{m^2}{2}\right) = 0.5$$

$$4m - \frac{m^2}{2} = 4$$

$$8m - m^2 = 8$$

$$m^2 - 8m + 8 = 0$$

Solve the quadratic equation by using the quadratic formula or by completing the square.

Using the quadratic formula:

$$m = \frac{-(-8) \pm \sqrt{(-8)^2 - 4 \times 1 \times 8}}{2 \times 1}$$

$$= \frac{8 \pm \sqrt{32}}{2}$$

So $m = \dfrac{8 + \sqrt{32}}{2} = 6.828\ldots$ or $m = \dfrac{8 - \sqrt{32}}{2} = 1.171\ldots$

Since m lies between 0 and 4, reject $m = 6.828\ldots$

So, median $= 1.17$ (3 s.f.)

Alternatively, by completing the square:

$$m^2 - 8m + 8 = 0$$

$$(m - 4)^2 - 16 + 8 = 0$$

$$(m - 4)^2 = 8$$

$$m - 4 = \pm\sqrt{8}$$

$$m = 4 \pm 2\sqrt{2}$$

$m = 4 + 2\sqrt{2} = 6.828...$(reject) or $m = 4 - 2\sqrt{2} = 1.171....$

So, median = 1.17 (3 s.f.), as above.

Note: As $y = f(x)$ is a straight line, the median could also be found using geometry.

Using geometry:

When $x = m$, $y = \frac{1}{8}(4 - m)$

Area of shaded triangle $= \frac{1}{2} \times (4 - m) \times \frac{1}{8}(4 - m)$

$$= \frac{1}{16}(4 - m)^2$$

But, since $P(X > m) = 0.5$, area of shaded triangle = 0.5.

So $\frac{1}{16}(4 - m)^2 = 0.5$

$$(4 - m)^2 = 8$$

Square root both sides:

$$4 - m = \pm\sqrt{8}$$

$m = 4 + \sqrt{8} = 6.828..$ (reject) or $m = 4 - \sqrt{8} = 1.171...$

So, median = 1.17 (3 s.f.), as above.

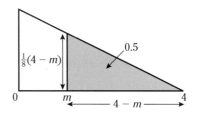

Exercise 3b

1 The continuous random variable X has probability density function given by

$$f(x) = \begin{cases} \frac{3}{8}x^2 & 0 \leqslant x \leqslant 2 \\ 0 & \text{otherwise} \end{cases}$$

   (i) Find the median.

   (ii) Find the upper and lower quartiles.

   (iii) Find the interquartile range.

2 The continuous random variable X has probability density function given by

$$f(x) = \begin{cases} 1 - \frac{1}{4}x & 1 \leqslant x \leqslant 3 \\ 0 & \text{otherwise} \end{cases}$$

   (i) Find the median.

   (ii) Given that X is greater than the median, find the probability that X is less than the upper quartile.

3 The continuous random variable X has probability density function given by

$$f(x) = \begin{cases} \frac{3}{2x^2} & 1 \leqslant x \leqslant 3 \\ 0 & \text{otherwise} \end{cases}$$

   (i) Find the median.

   (ii) Show that the upper quartile is 2 and find the lower quartile.

4 The continuous random variable X has probability density function given by

$$f(x) = \begin{cases} \frac{1 + x}{k} & 1 \leqslant x \leqslant 3 \\ 0 & \text{otherwise} \end{cases}$$

   where k is a constant.

   (i) Find the value of k.

   (ii) Sketch $y = f(x)$.

   (iii) Find the median.

(iv) Find the probability that exactly 4 out of 6 random observations of X have values less than the lower quartile.

5 The continuous random variable X has probability density function given by

$$f(x) = \begin{cases} \frac{1}{18}(3 + x) & -3 \le x \le 3 \\ 0 & \text{otherwise} \end{cases}$$

Find the lower quartile of X.

6 The continuous random variable X has probability density function given by

$$f(x) = \begin{cases} \frac{3}{8}(1 + x^2) & -1 \le x \le 1 \\ 0 & \text{otherwise} \end{cases}$$

(i) Sketch the probability density function.

(ii) **State** the value of the median.

7 The continuous random variable X has probability density function given by

$$f(x) = \begin{cases} 3x^{-2} & x \ge 3 \\ 0 & \text{otherwise} \end{cases}$$

(i) Find the median.

(ii) Show that $P(X > 12) = 0.25$ and hence state the value of the upper quartile.

(iii) Find the interquartile range.

8 The continuous random variable X has probability density function given by

$$f(x) = \begin{cases} \frac{32}{3}x^{-3} & 2 \le x \le 4 \\ 0 & \text{otherwise} \end{cases}$$

Find the median of X.

MEAN AND VARIANCE

Mean (expectation) of X

The mean of X, also called the expectation or expected value of X, is written $E(X)$ and is denoted by μ.

For a continuous random variable X with probability density function $f(x)$, the **mean** (or **expectation**) of X is given by

$$\mu = E(X) = \int_{\text{all } x} x f(x)\, dx$$

The formula for $E(X)$ is given in the examination.

Note: Compare this with discrete random variables, where $\mu = E(X) = \sum_{i=1}^{n} x_i p_i$ (S1 Chapter 4)

Example 3.8

The random variable X has probability density function given by

$$f(x) = \begin{cases} \frac{1}{18}(6 - x) & 0 \le x \le 6 \\ 0 & \text{otherwise} \end{cases}$$

Find $E(X)$.

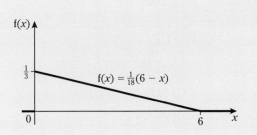

$$E(X) = \int_{\text{all } x} x f(x)\, dx$$

$$= \int_0^6 x \times \frac{1}{18}(6 - x)\, dx$$

Multiply x by $f(x)$ **before** integrating.

$$= \frac{1}{18} \int_0^6 (6x - x^2)\, dx$$

$$= \frac{1}{18} \left[3x^2 - \frac{x^3}{3} \right]_0^6$$

$$= \frac{1}{18} \left(\left(3 \times 6^2 - \frac{6^3}{3} \right) - 0 \right)$$

$$= 2$$

Example 3.9

The random variable X has probability density function given by

$$f(x) = \begin{cases} \frac{1}{9}x^2 & 0 \leqslant x \leqslant 3 \\ 0 & \text{otherwise} \end{cases}$$

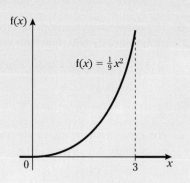

The mean of X is μ and the median of X is m.

(i) Find μ.

(ii) Find $P(X \leqslant \mu)$.

(iii) Find the probability that X lies between μ and m.

(i) $\mu = \displaystyle\int_{\text{all } x} x f(x)\,dx$

$\quad = \displaystyle\int_0^3 x \times \frac{1}{9} x^2 \,dx$

$\quad = \frac{1}{9} \displaystyle\int_0^3 x^3 \,dx$

$\quad = \frac{1}{9} \left[\dfrac{x^4}{4} \right]_0^3$

$\quad = \frac{1}{36} (3^4 - 0)$

$\quad = 2.25$

(ii) $P(X \leqslant \mu)$

$\quad = P(X \leqslant 2.25)$

$\quad = \displaystyle\int_0^{2.25} \frac{1}{9} x^2 \,dx$

$\quad = \frac{1}{9} \left[\dfrac{x^3}{3} \right]_0^{2.25}$

$\quad = \frac{1}{27} (2.25^3 - 0)$

$\quad = 0.421875$

$\quad = 0.422 \ (3 \text{ s.f.})$

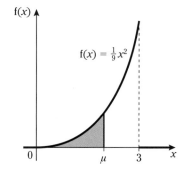

(iii) $P(X \leqslant m) = 0.5$ and $P(X \leqslant \mu) = 0.421875$, so $\mu < m$.

$\quad P(\mu \leqslant X \leqslant m) = 0.5 - 0.421875$

$\quad\quad\quad\quad\quad\quad = 0.078125$

$\quad\quad\quad\quad\quad\quad = 0.0781 \ (3 \text{ s.f.})$

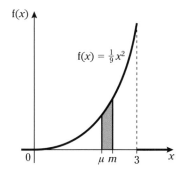

Example 3.10

The continuous random variable X has probability density function f, where $0 \leqslant x \leqslant 10$

The diagram shows the graph of $y = f(x)$.

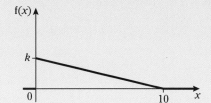

(i) Find the value of k.

(ii) Find E(X).

(i) The area under the graph is 1, so

$$\tfrac{1}{2} \times 10 \times k = 1$$

$$5k = 1$$

$$k = \tfrac{1}{5}$$

(ii) *You need to find the equation of the line.*

The general equation of a straight line is $y = mx + c$, where m is the gradient and c is the y-intercept.

Now $m = -\dfrac{\frac{1}{5}}{10} = -\tfrac{1}{50}$, $c = k = \tfrac{1}{5}$, so $y = -\tfrac{1}{50}x + \tfrac{1}{5}$

The probability density function of X is given by

$$f(x) = -\tfrac{1}{50}x + \tfrac{1}{5} \text{ for } 0 \leqslant x \leqslant 10$$

$$\begin{aligned}
E(X) &= \int_{\text{all } x} x f(x)\, dx \\
&= \int_0^{10} x\left(-\tfrac{1}{50}x + \tfrac{1}{5}\right) dx \\
&= \int_0^{10} \left(-\tfrac{1}{50}x^2 + \tfrac{1}{5}x\right) dx \\
&= \left[-\frac{x^3}{150} + \frac{x^2}{10}\right]_0^{10} \\
&= \left(-\frac{10^3}{150} + \frac{10^2}{10}\right) - 0 \\
&= 3\tfrac{1}{3}
\end{aligned}$$

Example 3.11

The continuous random variable T, has probability density function given by

$$f(t) = \begin{cases} \dfrac{k}{t^3} & t \geqslant 3 \\ 0 & \text{otherwise} \end{cases}$$

where k is a constant.

(i) Find the value of k.　　　　　　(ii) Find E(T).

(i) Total area under curve

$$= \int_{\text{all } t} f(t)\, dt$$

$$= \int_{3}^{\infty} \frac{k}{t^3}\, dt$$

The upper limit is infinity.

$$= k \int_{3}^{\infty} t^{-3}\, dt$$

$$= k \left[\frac{t^{-2}}{-2} \right]_{3}^{\infty}$$

$$= -\frac{k}{2} \left[\frac{1}{t^2} \right]_{3}^{\infty}$$

$$= -\frac{k}{2} \left(0 - \frac{1}{9} \right)$$

$\frac{1}{t^2} \to 0$ when $t \to \infty$.

$$= \frac{k}{18}$$

Total area $= 1$,

so $\quad \dfrac{k}{18} = 1$

$\quad\quad k = 18$

(ii) $f(t) = \dfrac{18}{t^3}$

$$E(T) = \int_{\text{all } t} t f(t)\, dt$$

$$= \int_{3}^{\infty} t \times \frac{18}{t^3}\, dt$$

$$= 18 \int_{3}^{\infty} \frac{1}{t^2}\, dt$$

$$= 18 \int_{3}^{\infty} t^{-2}\, dt$$

$$= 18 \left[\frac{t^{-1}}{-1} \right]_{3}^{\infty}$$

Useful integral: $\int \frac{1}{t^2} dt = -\frac{1}{t}$

$$= -18 \left[\frac{1}{t} \right]_{3}^{\infty}$$

$$= -18 \left(0 - \frac{1}{3} \right)$$

$\frac{1}{t} \to 0$ when $t \to \infty$.

$$= 6$$

Example 3.12

At a town centre car park the length of stay in hours is denoted by the random variable X, which has probability density function given by

$$f(x) = \begin{cases} kx^{-\frac{3}{2}} & 1 \leqslant x \leqslant 9, \\ 0 & \text{otherwise,} \end{cases}$$

where k is a constant.

   (i) Interpret the inequalities $1 \leqslant x \leqslant 9$ in the definition of $f(x)$ in the context of the question.
  (ii) Show that $k = \frac{3}{4}$.
 (iii) Calculate the mean length of stay.

The charge for a length of stay of x hours is $(1 - e^{-x})$ dollars.

(iv) Find the length of stay for a charge to be at least 0.75 dollars.

(v) Find the probability of the charge being at least 0.75 dollars.

Cambridge Paper 7 Q7 N06

(i) All cars stayed between 1 hour and 9 hours.

(ii) Total area under curve

$$= \int_{\text{all } x} f(x)\, dx$$

$$= \int_1^9 kx^{-\frac{3}{2}}\, dx$$

$$= k \left[\frac{x^{-\frac{1}{2}}}{-\frac{1}{2}} \right]_1^9$$

$$= -2k\left[x^{-\frac{1}{2}} \right]_1^9$$

$$= -2k\left(9^{-\frac{1}{2}} - 1^{-\frac{1}{2}} \right)$$

$$= -2k\left(\tfrac{1}{3} - 1 \right)$$

$$= \tfrac{4}{3}k$$

Total area $= 1$

so $\qquad \tfrac{4}{3}k = 1$

$$k = \tfrac{3}{4}$$

(iii) $f(x) = \tfrac{3}{4} x^{-\frac{3}{2}}$

$$\mu = \int_1^9 x \times \tfrac{3}{4} x^{-\frac{3}{2}}\, dx$$

$$= \tfrac{3}{4} \int_1^9 x^{-\frac{1}{2}}\, dx$$

$$= \tfrac{3}{4} \left[\frac{x^{\frac{1}{2}}}{\frac{1}{2}} \right]_1^9$$

$$= \tfrac{3}{2} \left[x^{\frac{1}{2}} \right]_1^9$$

$$= \tfrac{3}{2} \left(\sqrt{9} - \sqrt{1} \right)$$

$$= 3$$

(iv) Let t hours be the length of stay for the charge to be at least 0.75 dollars.

You require

$$1 - e^{-t} \geqslant 0.75$$

$$e^{-t} \leqslant 0.25$$

$$\frac{1}{e^t} \leqslant 0.25$$

$$e^t \geqslant 4$$

$$t \geqslant \ln 4$$

so $\qquad t \geqslant 1.386\ldots$ $\qquad\qquad$ Store 1.386... in your calculator memory.

i.e. $\qquad$ The length of stay is at least 1.39 hours (3 s.f.)

(v) If the <u>charge</u> is at least 0.75 dollars, then the <u>length of stay</u>, X, must be at least 1.386... hours.

$$P(X > 1.386...) = \frac{3}{4} \int_{1.386...}^{9} x^{-\frac{3}{2}} \, dx$$

You could write ln 4 instead of 1.386...

$$= \frac{3}{4} \left[\frac{x^{-\frac{1}{2}}}{-\frac{1}{2}} \right]_{1.386...}^{9}$$

$$= -\frac{3}{2} \left[x^{-\frac{1}{2}} \right]_{1.386...}^{9}$$

$$= -\frac{3}{2} \left(9^{-\frac{1}{2}} - (1.386...)^{-\frac{1}{2}} \right)$$

$$= 0.7739...$$

$$= 0.774 \text{ (3 s.f.)}$$

Example 3.13

The continuous random variable X has probability density function given by

$$f(x) = \begin{cases} ax + b & 0 \leqslant x \leqslant 2 \\ 0 & \text{otherwise} \end{cases}$$

where a and b are constants.

It is given that $E(X) = \frac{16}{15}$. Find the value of a and the value of b.

You have two unknowns to find, so you will need two equations.

First use the fact that the total area under the curve is 1.

Total area under curve

$$= \int_{\text{all } x} f(x) \, dx$$

$$= \int_{0}^{2} (ax + b) \, dx$$

$$= \left[\frac{ax^2}{2} + bx \right]_{0}^{2}$$

$$= \frac{a \times 2^2}{2} + b \times 2 - 0$$

$$= 2a + 2b$$

Total area $= 1$,

so $\quad 2a + 2b = 1 \qquad\qquad$ (1)

Now use the fact that $E(X) = \frac{16}{15}$

$$E(X) = \int_{\text{all } x} x f(x) \, dx$$

$$= \int_{0}^{2} x(ax + b) \, dx$$

$$= \int_{0}^{2} (ax^2 + bx) \, dx$$

$$= \left[\frac{ax^3}{3} + \frac{bx^2}{2} \right]_0^2$$

$$= \frac{a \times 2^3}{3} + \frac{b \times 2^2}{2} - 0$$

$$= \frac{8a}{3} + 2b$$

So, $\frac{8a}{3} + 2b = \frac{16}{15}$

Multiply through by 15:

$\quad$ $40a + 30b = 16$

Divide through by 2:

$\quad$ $20a + 15b = 8$ $\qquad\qquad\qquad$ (2)

Now solve the simultaneous equations.

$\quad$ (1) $\times$ 10 $\qquad$ $20a + 20b = 10$ $\qquad$ (3)

$\quad$ (2) $\qquad\qquad$ $20a + 15b = 8$ $\qquad$ (2)

$\quad$ (3) $-$ (2) $\qquad\qquad\quad$ $5b = 2$

$\qquad\qquad\qquad\qquad\qquad$ $b = \frac{2}{5}$

Substitute into (1)

$\quad$ $2a + 2 \times \frac{2}{5} = 1$

$\qquad$ $2a = \frac{1}{5}$

$\qquad\quad$ $a = \frac{1}{10}$

So, $a = \frac{1}{10}$ and $b = \frac{2}{5}$

Note about symmetry

If the probability density function f is defined for $a \leqslant x \leqslant b$ and the graph of $y = f(x)$ has a line of symmetry in this interval, then the **mean** is the **midpoint of the interval:**

$$\mu = E(X) = \tfrac{1}{2}(a + b)$$

For example, consider the random variable X defined in Example 3.2.

$$f(x) = \begin{cases} \frac{1}{36}x(6 - x) & 0 \leqslant x \leqslant 6 \\ 0 & \text{otherwise} \end{cases}$$

The graph of $y = f(x)$ is part of a quadratic curve.

Since $y = 0$ when $x = 0$ and $x = 6$, the curve is symmetrical about the line $x = 3$.

So, by symmetry,

$$\mu = E(X) = 3$$

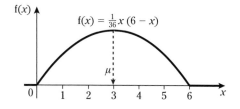

Check by integration:

$$\mu = \text{E}(X) = \int_{\text{all } x} x\text{f}(x)\,dx$$

$$= \int_0^6 x \times \tfrac{1}{36} x(6 - x)\,dx$$

$$= \tfrac{1}{36} \int_0^6 (6x^2 - x^3)\,dx$$

$$= \tfrac{1}{36} \left[2x^3 - \frac{x^4}{4} \right]_0^6$$

$$= \tfrac{1}{36} \left(2 \times 6^3 - \frac{6^4}{4} - 0 \right)$$

$$= 3, \text{ as expected.}$$

Always check for symmetry, as this could save you time.

Exercise 3c

1 The continuous random variable X has probability density function given by

$$\text{f}(x) = \begin{cases} \tfrac{1}{16} x & 2 \leqslant x \leqslant 6 \\ 0 & \text{otherwise} \end{cases}$$

Find $\text{E}(X)$.

2 The continuous random variable X has probability density function given by

$$\text{f}(x) = \begin{cases} \dfrac{x + 3}{20} & 0 \leqslant x \leqslant 4 \\ 0 & \text{otherwise} \end{cases}$$

Find the value of $\text{E}(X)$.

3 The continuous random variable X has probability density function given by

$$\text{f}(x) = \begin{cases} \tfrac{3}{4}(x^2 + 1) & 0 \leqslant x \leqslant 1 \\ 0 & \text{otherwise} \end{cases}$$

Find the value of $\text{E}(X)$.

4 The continuous random variable Y has probability density function given by

$$\text{f}(y) = \begin{cases} \tfrac{3}{14}\sqrt{y} & 1 \leqslant y \leqslant 4 \\ 0 & \text{otherwise} \end{cases}$$

Find the value of $\text{E}(Y)$.

5 The continuous random variable X has probability density function given by

$$\text{f}(x) = \begin{cases} \tfrac{3}{4}x(2 - x) & 0 \leqslant x \leqslant 2 \\ 0 & \text{otherwise} \end{cases}$$

(i) Sketch the probability density function of X.

(ii) Find the mean value of X.

6 The continuous random variable X has probability density function given by

$$\text{f}(x) = \begin{cases} \tfrac{1}{4}x^3 & 0 \leqslant x \leqslant 2 \\ 0 & \text{otherwise} \end{cases}$$

(i) Find $\text{E}(X)$.

(ii) Find $\text{P}(X < \text{E}(X))$.

(iii) Is the mean of X less than or greater than the median of X? Justify your answer.

7 The random variable X denotes the lifetime, in years, of a particular type of light bulb. The probability density function of X is given by

$$\text{f}(x) = \begin{cases} kx(5 - x) & 0 \leqslant x \leqslant 5 \\ 0 & \text{otherwise} \end{cases}$$

(i) Show that $k = \tfrac{6}{125}$

(ii) Two light bulbs are selected at random. Find the probability that both light bulbs last longer than the mean lifetime of this type of light bulb.

8 The continuous random variable X has probability density function given by

$$\text{f}(x) = \begin{cases} \tfrac{5}{32}x^4 & 0 \leqslant x \leqslant 2 \\ 0 & \text{otherwise} \end{cases}$$

(i) Find $\text{E}(X)$.

(ii) Find the median m.

(iii) Find the probability that a random observation of X lies between the mean and the median.

9 The continuous random variable X has probability density function given by

$$f(x) = \begin{cases} k & a \leq x \leq b \\ 0 & \text{otherwise} \end{cases}$$

where k, a and b are positive constants.

(i) Express k in terms of a and b.

The mean of X is 8 and the interquartile range is 6.

(ii) Find the values of a, b and k.

(iii) Find $P(X > 12)$.

10 The continuous random variable T has probability density function given by

$$f(t) = \begin{cases} \dfrac{k}{t^4} & t \geq 1 \\ 0 & \text{otherwise} \end{cases}$$

where k is a constant.

(i) Show that $k = 3$.

(ii) Find $E(T)$.

11 The continuous random variable X has probability density function given by

$$f(x) = \begin{cases} \dfrac{k}{\sqrt[3]{x}} & 1 \leq x \leq 8 \\ 0 & \text{otherwise} \end{cases}$$

where k is a constant.

(i) Show that $k = \frac{2}{9}$.

(ii) Find $E(X)$.

12 The continuous random variable X has probability density function given by

$$f(x) = \begin{cases} p - qx & 0 \leq x \leq 2 \\ 0 & \text{otherwise} \end{cases}$$

where p and q are constants.

(i) Show that $2p - 2q = 1$.

(ii) Given that the mean of X is $\frac{2}{3}$,

(a) form a second equation in p and q,

(b) find the value of p and the value of q.

Variance of X

For continuous random variables, the variance of X, denoted by σ^2, is defined as follows:

$$\sigma^2 = \text{Var}(X) = \int_{\text{all } x} (x - \mu)^2 \, f(x) \, dx \qquad \text{where } \mu = E(X)$$

However, this formula can be complicated to work with, so an alternative version derived by expanding the bracket is usually used. This is shown below.

For a continuous random variable X with probability density function f, the **variance** of X, $\text{Var}(X)$, is denoted by σ^2, where

$$\sigma^2 = \text{Var}(X) = \int_{\text{all } x} x^2 \, f(x) \, dx - \mu^2 \qquad \text{where } \mu = \int_{\text{all } x} xf(x) \, dx$$

This compares with the two versions of the variance formula for discrete random variables where

$$\text{Var}(X) = \sum_{i=1}^{n} (x_i - \mu)^2 \, p_i = \sum_{i=1}^{n} x_i^2 \, p_i - \mu^2$$

On the formulae list provided in the examination, the expectation and variance formulae are given as follows:

$$E(X) = \int xf(x) \, dx \qquad \text{Var}(X) = \int x^2 \, f(x) \, dx - \{E(X)\}^2$$

Example 3.14

The random variable X has probability density function given by

$$f(x) = \begin{cases} 3x^k & 0 \leqslant x \leqslant 1 \\ 0 & \text{otherwise} \end{cases}$$

where k is a positive constant.

(i) Find the value of k.

(ii) Show that the mean, μ, of X is 0.75

(iii) Show that the standard deviation, σ, is 0.1936, correct to 4 significant figures.

(iv) Find $P(\mu - \sigma \leqslant X \leqslant \mu + \sigma)$.

(i) Total area under curve

$$= \int_{\text{all } x} f(x)\,dx$$

$$= \int_0^1 3x^k\,dx$$

$$= 3\int_0^1 x^k\,dx$$

$$= 3\left[\frac{x^{k+1}}{k+1}\right]_0^1$$

$$= \frac{3}{k+1}\left[x^{k+1}\right]_0^1$$

$$= \frac{3}{k+1}(1-0) \qquad\qquad 1^{k+1} = 1$$

$$= \frac{3}{k+1}$$

Total area $= 1$

so $\quad \dfrac{3}{k+1} = 1$

$$k + 1 = 3$$

$$k = 2$$

(ii) $f(x) = 3x^2$

$$\mu = \int_{\text{all } x} xf(x)\,dx$$

$$= \int_0^1 x \times 3x^2\,dx$$

$$= 3\int_0^1 x^3\,dx$$

$$= 3\left[\frac{x^4}{4}\right]_0^1$$

$$= \tfrac{3}{4}(1-0)$$

$$= 0.75$$

(iii) *First find the variance of X.*

$$\sigma^2 = \int_{\text{all } x} x^2 \text{f}(x)\, dx - \mu^2$$

$$= \int_0^1 x^2 \times 3x^2\, dx - \mu^2$$

$$= 3 \int_0^1 x^4\, dx - \mu^2$$

$$= 3 \left[\frac{x^5}{5} \right]_0^1 - \mu^2$$

$$= 3 \left(\frac{1^5}{5} - 0 \right) - (0.75)^2$$

$$= 0.0375$$

$$\sigma = \sqrt{0.0375} = 0.19364\ldots = 0.1936 \text{ (4 s.f.)}$$

(iv) $\text{P}(\mu - \sigma \leqslant X \leqslant \mu + \sigma)$

$$= \int_{\mu - \sigma}^{\mu + \sigma} 3x^2\, dx$$

$$= \left[x^3 \right]_{\mu - \sigma}^{\mu + \sigma}$$

$$= (0.9436\ldots)^3 - (0.5563\ldots)^3$$

$$= 0.6680\ldots$$

$$= 0.668 \text{ (3 s.f.)}$$

$\mu + \sigma = 0.75 + \sqrt{0.0375} = 0.9436\ldots$

$\mu - \sigma = 0.75 - \sqrt{0.0375} = 0.5563\ldots$

Note that it is better to use $\sigma = \sqrt{0.0375}$ here but, if you do use a rounded value, you must use at least 4 significant figures.

Example 3.15

A continuous random variable X has probability density function given by

$$\text{f}(x) = \begin{cases} 3(1 - x)^2 & 0 \leqslant x \leqslant 1, \\ 0 & \text{otherwise.} \end{cases}$$

Find:

(i) $\text{P}(X > 0.5)$.

(ii) The mean and variance of X. Cambridge Paper 7 Q6 N04

(i) $\text{P}(X > 0.5) = \int_{0.5}^1 \text{f}(x)\, dx$

$$= \int_{0.5}^1 3(1 - x)^2\, dx$$

$$= 3 \int_{0.5}^1 (1 - 2x + x^2)\, dx$$

$$= 3 \left[x - x^2 + \frac{x^3}{3} \right]_{0.5}^1$$

$$= 3 \left(\left(1 - 1 + \frac{1}{3} \right) - \left(0.5 - 0.5^2 + \frac{0.5^3}{3} \right) \right)$$

$$= 0.125$$

Alternatively,

$$P(X > 0.5) = \int_{0.5}^{1} 3(1-x)^2 \, dx$$

$$= -\left[(1-x)^3\right]_{0.5}^{1}$$

$$= -(0 - 0.5^3)$$

$$= 0.125$$

> You may have studied this direct method of integrating $(ax + b)^n$ where $n \neq -1$. Alternatively, the substitution $u = 1 - x$ can be used.

(ii)

$$\mu = \int_{\text{all } x} x f(x) \, dx$$

$$= \int_0^1 x \times 3(1-x)^2 \, dx$$

$$= 3\int_0^1 x(1 - 2x + x^2) \, dx$$

$$= 3\int_0^1 (x - 2x^2 + x^3) \, dx$$

$$= 3\left[\frac{x^2}{2} - \frac{2x^3}{3} + \frac{x^4}{4}\right]_0^1$$

$$= 3\left(\frac{1}{2} - \frac{2}{3} + \frac{1}{4} - 0\right)$$

$$= \frac{1}{4}$$

$$\text{Var}(X) = \int_{\text{all } x} x^2 f(x) \, dx - \mu^2$$

$$= \int_0^1 x^2 \times 3(1-x)^2 \, dx - \mu^2$$

$$= 3\int_0^1 x^2(1 - 2x + x^2) \, dx - \mu^2$$

$$= 3\int_0^1 (x^2 - 2x^3 + x^4) \, dx - \mu^2$$

$$= 3\left[\frac{x^3}{3} - \frac{2x^4}{4} + \frac{x^5}{5}\right]_0^1 - \left(\frac{1}{4}\right)^2$$

$$= 3\left(\frac{1}{3} - \frac{1}{2} + \frac{1}{5} - 0\right) - \frac{1}{16}$$

$$= \frac{3}{80}$$

Example 3.16

A continuous random variable X has probability density function given by

$$f(x) = \begin{cases} \dfrac{1}{c} & 0 \leqslant x \leqslant c \\ 0 & \text{otherwise} \end{cases}$$

(i) State the value of $E(X)$ in terms of c.

(ii) Find $\text{Var}(X)$ in terms of c.

(iii) If $c = 6$, find the standard deviation of X.

(i) *It is useful to draw a sketch.*

By symmetry,

$E(X) = \mu = \frac{1}{2}c$

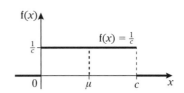

(ii) $Var(X) = \int_{all\ x} x^2 f(x)\, dx - \mu^2$

$= \int_0^c x^2 \times \frac{1}{c}\, dx - \mu^2$

$= \frac{1}{c} \left[\frac{x^3}{3} \right]_0^c - \mu^2$

$= \frac{1}{3c}(c^3 - 0) - \left(\frac{1}{2}c \right)^2$

$= \frac{1}{3}c^2 - \frac{1}{4}c^2$

$= \frac{1}{12}c^2$

(iii) When $c = 6$, $Var(X) = \frac{1}{12} \times 6^2 = 3$

So, standard deviation $= \sqrt{3} = 1.732... = 1.73$ (3 s.f.)

Exercise 3d

1 The continuous random variable X has probability density function given by

$$f(x) = \begin{cases} \frac{2}{5}x & 2 \leqslant x \leqslant 3 \\ 0 & \text{otherwise} \end{cases}$$

(i) Find $E(X)$.

(ii) Find $Var(X)$.

2 The continuous random variable X has probability density function given by

$$f(x) = \begin{cases} \frac{3}{7}x^2 & 1 \leqslant x \leqslant 2 \\ 0 & \text{otherwise} \end{cases}$$

(i) Find $E(X)$.

(ii) Find $Var(X)$.

(iii) Find the standard deviation of X.

3 The continuous random variable X has probability density function given by

$$f(x) = \begin{cases} k & -2 \leqslant x \leqslant 3 \\ 0 & \text{otherwise} \end{cases}$$

(i) Find the value of k.

(ii) **State** the value of $E(X)$.

(iii) Find $Var(X)$.

(iv) Find the standard deviation of X.

4 The continuous random variable X has mean μ and standard deviation σ. The probability density function of X is given by

$$f(x) = \begin{cases} \frac{1}{32}(8 - x) & 0 \leqslant x \leqslant 8 \\ 0 & \text{otherwise} \end{cases}$$

(i) Show that $\mu = 2\frac{2}{3}$

(ii) Show that $\sigma^2 = 3\frac{5}{9}$

(iii) Find $P(X \geqslant \mu + \sigma)$.

5 The continuous random variable X has probability density function given by

$$f(x) = \begin{cases} \frac{3}{16}(4 - x^2) & 0 \leqslant x \leqslant 2 \\ 0 & \text{otherwise} \end{cases}$$

(i) Show that the mean of X is 0.75

(ii) Find the variance of X.

6 The continuous random variable T has probability density function given by

$$f(t) = \begin{cases} \frac{5}{6}(t^4 + 1) & 0 \leqslant t \leqslant 1 \\ 0 & \text{otherwise} \end{cases}$$

(i) Show that $E(T) = \frac{5}{9}$

(ii) Find the variance of T.

7 The mass, in kilograms, of metal extracted from 10 g of ore from a certain mine is a continuous random variable X with probability density function

$$f(x) = \begin{cases} \frac{3}{4}x(2-x)^2 & 0 \leq x \leq 2 \\ 0 & \text{otherwise} \end{cases}$$

(i) Show that the mean mass is 0.8 kg.

(ii) Find the standard deviation of the mass of metal extracted.

8 The continuous random variable T has probability density function given by

$$f(t) = \begin{cases} \dfrac{k}{\sqrt{t}} & 1 \leq t \leq 4 \\ 0 & \text{otherwise} \end{cases}$$

where k is a constant.

(i) Find the value of k.

(ii) Find the standard deviation of T.

9 The continuous random variable Y has probability density function given by

$$f(y) = \begin{cases} \dfrac{a}{y^4} & y \geq 2 \\ 0 & \text{otherwise} \end{cases}$$

where a is a constant.

(i) Show that $a = 24$.

(ii) Find $E(Y)$.

(iii) Find $Var(Y)$.

MISCELLANEOUS WORKED EXAMPLES

The examples in this section include P3 integration methods.

Example 3.17

The random variable T denotes the time in seconds for which a firework burns before exploding. The probability density function for T is given by

$$f(t) = \begin{cases} ke^{0.2t} & 0 \leq t \leq 5, \\ 0 & \text{otherwise}, \end{cases}$$

where k is a constant.

(i) Show that $k = \dfrac{1}{5(e-1)}$.

(ii) Sketch the probability density function.

(iii) 80% of fireworks burn for longer than a certain time before they explode. Find this time.

Cambridge Paper 7 Q5 J10(71)

(i) Total area under curve

$$= \int_{\text{all } t} f(t)\, dt$$

$$= \int_0^5 ke^{0.2t}\, dt$$

$$= k\left[\frac{e^{0.2t}}{0.2}\right]_0^5$$

$$= \frac{k}{0.2}(e^1 - e^0) \qquad \frac{1}{0.2} = 5,\ e^1 = e,\ e^0 = 1$$

$$= 5k(e-1)$$

Total area $= 1$

so $\quad 5k(e-1) = 1$

$$k = \frac{1}{5(e-1)}$$

(ii) The curve is exponential.

When $t = 0$, $y = k\mathrm{e}^0 = k \times 1 = \dfrac{1}{5(\mathrm{e} - 1)} = 0.116...$

When $t = 5$, $y = k\mathrm{e}^{0.2 \times 5} = k\mathrm{e}^1 = \dfrac{\mathrm{e}}{5(\mathrm{e} - 1)} = 0.316...$

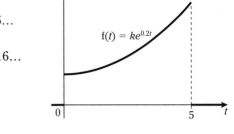

(iii) $P(T > t_1) = \displaystyle\int_{t_1}^{5} k\mathrm{e}^{0.2t}\,\mathrm{d}t$

$\qquad\quad = k\left[\dfrac{\mathrm{e}^{0.2t}}{0.2}\right]_{t_1}^{5}$

$\qquad\quad = 5 \times \dfrac{1}{5(\mathrm{e} - 1)}\,(\mathrm{e} - \mathrm{e}^{0.2t_1})$

$\qquad\quad = \dfrac{\mathrm{e} - \mathrm{e}^{0.2t_1}}{\mathrm{e} - 1}$

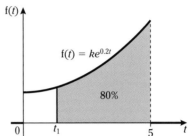

But you are given that $P(T > t_1) = 0.8$

so $\quad \dfrac{\mathrm{e} - \mathrm{e}^{0.2t_1}}{\mathrm{e} - 1} = 0.8$

$\qquad \mathrm{e} - \mathrm{e}^{0.2t_1} = 0.8(\mathrm{e} - 1)$

$\quad \mathrm{e} - 0.8\mathrm{e} + 0.8 = \mathrm{e}^{0.2t_1}$

$\qquad\quad \mathrm{e}^{0.2t_1} = 0.2\mathrm{e} + 0.8$

$\qquad\qquad\quad = 1.3436...$

Take $\log_e$ of both sides.

$\qquad 0.2t_1 = \ln(1.3436...) = 0.2953...$

$\qquad\quad t_1 = 5 \times 0.2953...$

$\qquad\qquad = 1.4769...$

$\qquad\qquad = 1.48\ (3\ \text{s.f.})$

So 80% of fireworks burn for longer than $1.48\,\text{s}$.

Example 3.18

The continuous random variable X has probability density function given by

$$f(x) = \begin{cases} k\cos x & 0 \leqslant x \leqslant \tfrac{1}{4}\pi, \\ 0 & \text{otherwise,} \end{cases}$$

where k is a constant.

   (i) Show that $k = \sqrt{2}$.

  (ii) Find $P(X > 0.4)$.

 (iii) Find the upper quartile of X.

 (iv) Find the probability that exactly 3 out of 5 random observations of X have values greater than the upper quartile.

Cambridge Paper 7 Q5 N09(71)

(i) Total area under curve

$$= \int_{\text{all } x} \text{f}(x)\,dx$$

$$= \int_0^{\frac{1}{4}\pi} k\cos x\,dx$$

$$= k\left[\sin x\right]_0^{\frac{1}{4}\pi}$$

$$= k\left(\sin \tfrac{1}{4}\pi - \sin 0\right)$$

$$= k\left(\frac{1}{\sqrt{2}} - 0\right)$$

$$= \frac{k}{\sqrt{2}}$$

Give the **exact** value of $\sin \tfrac{1}{4}\pi$ using surds.

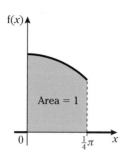

Total area $= 1$

so $\dfrac{k}{\sqrt{2}} = 1$

$k = \sqrt{2}$

(ii) $P(X > 0.4) = \displaystyle\int_{0.4}^{\frac{1}{4}\pi} \sqrt{2}\cos x\,dx$

$$= \sqrt{2}\left[\sin x\right]_{0.4}^{\frac{1}{4}\pi}$$

$$= \sqrt{2}\left(\frac{1}{\sqrt{2}} - \sin(0.4)\right)$$

$$= 0.4492\ldots$$

$$= 0.449 \ (3 \text{ s.f.})$$

You **must** work in radians, so set your calculator to RAD mode.

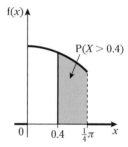

(iii) Denoting the upper quartile by q_3,

$$P(X \leqslant q_3) = \int_0^{q_3} \sqrt{2}\cos x\,dx$$

$$= \sqrt{2}\left[\sin x\right]_0^{q_3}$$

$$= \sqrt{2}\left(\sin q_3 - \sin 0\right)$$

$$= \sqrt{2}\sin q_3$$

But $P(X \leqslant q_3) = 0.75$

so $\sqrt{2}\sin q_3 = 0.75$

$$\sin q_3 = \frac{0.75}{\sqrt{2}} = 0.5303\ldots$$

$$q_3 = \sin^{-1}(0.5303\ldots)$$

$$= 0.5589\ldots$$

$$= 0.559 \text{ radians } (3 \text{ s.f.})$$

(iv) $P(X > q_3) = 0.25$

Let Y be the number of observations out of 5 random observations that are greater than q_3.

$Y \sim \text{B}(5, 0.25)$ $n = 5, p = 0.25, q = 1 - p = 0.75$

$$P(Y = 3) = \binom{5}{3} \times 0.25^3 \times 0.75^2$$

Recall: Binomial distribution (S1 Chapter 5)

number of ways to   probability of   probability of
choose 3 from 5   3 successes   2 failures

$$= 0.08789\ldots$$

$$= 0.0879 \ (3 \text{ s.f.})$$

Note: You do not need to know the value of the upper quartile in order to do part (iv).

Example 3.19

If Usha is stung by a bee she always develops an allergic reaction. The time taken in minutes for Usha to develop the reaction can be modelled using the probability density function given by

$$f(t) = \begin{cases} \dfrac{k}{t+1} & 0 \leq t \leq 4, \\ 0 & \text{otherwise,} \end{cases}$$

where k is a constant.

(i) Show that $k = \dfrac{1}{\ln 5}$.

(ii) Find the probability that it takes more than 3 minutes for Usha to develop the reaction.

(iii) Find the median time for Usha to develop a reaction.

Cambridge Paper 7 Q7 J08

(i) Total area under curve

$$= \int_{\text{all } t} f(t)\, dt$$

$$= \int_{0}^{4} \frac{k}{t+1}\, dt$$

$$= k\big[\ln(t+1)\big]_{0}^{4}$$

$$= k(\ln 5 - \ln 1)$$

$$= k \ln 5 \qquad \text{Recall: } \ln 1 = 0$$

Total area $= 1$

so $k \ln 5 = 1$

$$k = \frac{1}{\ln 5}$$

(ii) $P(T > 3) = \displaystyle\int_{3}^{4} \frac{k}{t+1}\, dt$

$$= k\big[\ln(t+1)\big]_{3}^{4}$$

$$= k(\ln 5 - \ln 4)$$

$$= \frac{1}{\ln 5}(\ln 5 - \ln 4)$$

$$= 0.1386\ldots$$

$$= 0.139 \ (3 \text{ s.f.})$$

Alternatively,

$$P(T > 3) = 1 - P(T \leq 3)$$

$$= 1 - \int_{0}^{3} \frac{k}{t+1}\, dt$$

$$= 1 - k\big[\ln(t+1)\big]_{0}^{3}$$

$$= 1 - k(\ln 4 - \ln 1)$$

$$= 1 - \frac{1}{\ln 5} \times \ln 4$$

$$= 0.1386\ldots$$

(iii) Denoting the median by m,

$$P(T \leqslant m) = \int_0^m \frac{k}{t+1} \, dt$$

$$= k\left[\ln(t+1)\right]_0^m$$

$$= k(\ln(m+1) - \ln 1)$$

$$= \frac{\ln(m+1)}{\ln 5}$$

But $P(X \leqslant m) = 0.5$

so $\dfrac{\ln(m+1)}{\ln 5} = 0.5$

$$\ln(m+1) = 0.5 \times \ln 5$$

$$= 0.8047\ldots$$

$$m + 1 = e^{0.8047\ldots}$$

$$= 2.236\ldots$$

$$m = 1.236\ldots$$

$$= 1.24 \text{ (3 s.f.)}$$

The median time for Usha to develop the reaction is 1.24 min.

Example 3.20

The lifetime, x years, of the power light on a freezer, which is left on continuously, can be modelled by the continuous random variable with density function given by

$$f(x) = \begin{cases} ke^{-3x} & x > 0, \\ 0 & \text{otherwise,} \end{cases}$$

where k is a constant.

(i) Show that $k = 3$.

(ii) Find the lower quartile.

(iii) Find the mean lifetime.

Cambridge Paper 7 Q7 N03

(i) Total area under curve

$$= \int_{\text{all } x} f(x) \, dx$$

$$= \int_0^\infty ke^{-3x} \, dx$$

$$= k\left[\frac{e^{-3x}}{-3}\right]_0^\infty \qquad\qquad e^{-3x} \to 0 \text{ as } x \to \infty$$

$$= -\frac{k}{3}(0 - e^0)$$

$$= -\frac{k}{3}(-1)$$

$$= \frac{k}{3}$$

Total area $= 1$

so $\dfrac{k}{3} = 1$

$$k = 3$$

(ii) For the lower quartile q_1,

$$P(X \leqslant q_1) = \int_0^{q_1} 3e^{-3x}\,dx$$

$$= \left[-e^{-3x}\right]_0^{q_1}$$

$$= -e^{-3q_1} - (-e^0)$$

$$= -e^{-3q_1} + 1$$

But $P(X \leqslant q_1) = 0.25$

so $0.25 = -e^{-3q_1} + 1$

$$e^{-3q_1} = 0.75$$

$$-3q_1 = \ln(0.75)$$

$$q_1 = \frac{\ln(0.75)}{-3} = 0.09589\ldots = 0.0959 \text{ (3 s.f.)}$$

The lower quartile is 0.0959 years.

(iii) $E(X) = \int_{\text{all } x} x f(x)\,dx$

$$= \int_0^\infty x \times 3e^{-3x}\,dx$$

$$= \left[x \times (-e^{-3x})\right]_0^\infty - \int_0^\infty (-e^{-3x}) \times 1\,dx \qquad \text{Use integration by parts.}$$

$$= \left[-xe^{-3x}\right]_0^\infty + \int_0^\infty e^{-3x}\,dx$$

$$= 0 + \left[\frac{e^{-3x}}{-3}\right]_0^\infty \qquad\qquad xe^{-3x} \to 0 \text{ as } x \to \infty, \text{ so } \left[-xe^{-3x}\right]_0^\infty = 0$$

$$= -\frac{1}{3}\left[e^{-3x}\right]_0^\infty$$

$$= -\frac{1}{3}(0 - e^0) = \frac{1}{3} \qquad\qquad e^{-3x} \to 0 \text{ as } x \to \infty$$

The mean lifetime $= \frac{1}{3}$ years $= 4$ months.

Exercise 3e

1 The continuous random variable X has probability density function given by

$$f(x) = \begin{cases} \frac{32}{3}x^{-3} & 2 \leqslant x \leqslant 4 \\ 0 & \text{otherwise} \end{cases}$$

(i) Find $E(X)$.   (ii) Find $\text{Var}(X)$.

2 The random variable T has probability density function given by

$$f(t) = \begin{cases} ke^t & 0 \leqslant t \leqslant 1 \\ 0 & \text{otherwise} \end{cases}$$

where k is a constant.

(i) Show that $k = \dfrac{1}{e - 1}$

(ii) Find the median of T.

(iii) Find the mean of T.

3 The continuous random variable X has probability density function given by

$$f(x) = \begin{cases} k \sin x & 0 \leqslant x \leqslant \pi \\ 0 & \text{otherwise} \end{cases}$$

where k is a constant.

(i) Show that $k = \frac{1}{2}$ Remember to work in radians.

(ii) By considering a sketch of the probability density function, state the value of the mean of X.

(iii) (a) Show that $P\left(X < \frac{1}{3}\pi\right) = \frac{1}{4}$

(b) Find $P\left(X < \frac{2}{3}\pi\right)$.

(c) Find the interquartile range.

4 The continuous random variable X has probability density function given by

$$f(x) = \begin{cases} k(x-1)^6 & 1 \le x \le 2 \\ 0 & \text{otherwise} \end{cases}$$

where k is a constant.

Using the substitution $u = x - 1$:

(i) show that $k = 7$

(ii) find the mean of X

(iii) find the median of X.

5 The continuous random variable X has probability density function given by

$$f(x) = \begin{cases} k \sec^2 x & 0 \le x \le \frac{1}{4}\pi \\ 0 & \text{otherwise} \end{cases}$$

where k is a constant.

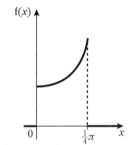

(i) Show that $k = 1$

Remember to work in radians.

(ii) Find the median of X.

(iii) Find the interquartile range.

6 The continuous random variable X has probability density function given by

$$f(x) = \begin{cases} \dfrac{k}{x} & 1 \le x \le 4 \\ 0 & \text{otherwise} \end{cases}$$

where k is a constant.

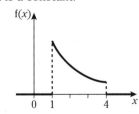

(i) Show that $k = \dfrac{1}{2 \ln 2}$

(ii) Find $E(X)$.

(iii) Find $Var(X)$.

(iv) Show that the median is 2.

(v) Show that the lower quartile is $\sqrt{2}$ and find the upper quartile.

7 The time, in years, that Eduardo keeps his car before replacing it with a new one can be modelled by a continuous random variable T with probability density function given by

$$f(t) = \begin{cases} \frac{1}{4}e^{-\frac{1}{4}t} & t > 0 \\ 0 & \text{otherwise} \end{cases}$$

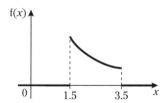

(i) Find the probability that he keeps his car less than 1 year before replacing it.

(ii) Find the probability that he keeps his car for more than 2 years before replacing it.

(iii) Find the mean length of time Eduardo keeps his car before replacing it.

8 The continuous random variable X has probability density function given by

$$f(x) = \begin{cases} \dfrac{kx^2}{x^3 - 1} & 1.5 \le x \le 3.5 \\ 0 & \text{otherwise} \end{cases}$$

where k is a constant.

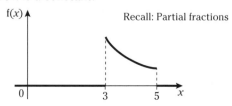

Find the value of k.

9 The continuous random variable X has probability density function given by

$$f(x) = \begin{cases} \dfrac{k}{(x-1)(x-2)} & 3 \le x \le 5 \\ 0 & \text{otherwise} \end{cases}$$

where k is a constant.

Recall: Partial fractions

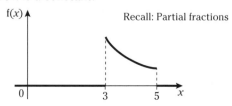

Find the value of k.

10 The continuous random variable X has probability density function given by

$$f(x) = \begin{cases} kxe^{2x} & 0 \leqslant x \leqslant 0.5 \\ 0 & \text{otherwise} \end{cases}$$

where k is a constant.

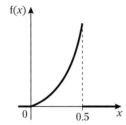

(i) Find the value of k.

(ii) Show that $E(X) = \frac{1}{2}e - 1$.

Recall: Integration by parts

11 The continuous random variable T has probability density function given by

$$f(t) = \begin{cases} \dfrac{k}{t+1} & 1 \leqslant t \leqslant 3 \\ 0 & \text{otherwise} \end{cases}$$

where k is a constant.

(i) Show that $k = \dfrac{1}{\ln 2}$

(ii) By using the substitution $u = t + 1$, or otherwise, find $E(T)$.

Summary

For a **continuous random variable** X with probability density function defined by $f(x)$ for $a \leqslant x \leqslant b$:

Probabilities are given by areas under the curve

$$P(x_1 \leqslant X \leqslant x_2) = \int_{x_1}^{x_2} f(x)\, dx$$

The total area under the curve is 1

$$\int_{\text{all } x} f(x)\, dx = 1$$

$f(x) \geqslant 0$ for all values of x, so the graph of $y = f(x)$ never goes below the x-axis.

Median and quartiles

Median m

$$P(X \leqslant m) = \int_a^m f(x)\, dx = 0.5$$

Lower quartile q_1

$$P(X \leqslant q_1) = \int_a^{q_1} f(x)\, dx = 0.25$$

Upper quartile q_3

$$P(X \leqslant q_3) = \int_a^{q_3} f(x)\, dx = 0.75$$

Mean and variance

Mean

$$\mu = E(X) = \int_{\text{all } x} x f(x)\, dx$$

Variance

$$\sigma^2 = \text{Var}(X) = \int_{\text{all } x} x^2 f(x)\, dx - \mu^2$$

Mixed Exercise 3

1 The random variable X has probability density function given by

$$f(x) = \begin{cases} kx & 0 \leqslant x \leqslant 2 \\ 0 & \text{otherwise} \end{cases}$$

where k is a constant.

  (i) Find the value of k.

  (ii) Find the median of X.

  (iii) Find the mean of X.

2 The random variable X has probability density function given by

$$f(x) = \begin{cases} 4x^k & 0 \leqslant x \leqslant 1 \\ 0 & \text{otherwise} \end{cases}$$

where k is a positive constant.

  (i) Show that $k = 3$

  (ii) Show that the mean of X is 0.8 and find the variance of X.

  (iii) Find the upper quartile of X.

  (iv) Find the interquartile range of X.
         Cambridge Paper 7 Q5 J06

3 A continuous random variable X has probability density function given by

$$f(x) = \begin{cases} \frac{1}{6}x & 2 \leqslant x \leqslant 4, \\ 0 & \text{otherwise.} \end{cases}$$

  (i) Find $E(X)$.

  (ii) Find the median of X.

  (iii) Two independent values of X are chosen at random. Find the probability that both these values are greater than 3.
         Cambridge Paper 7 Q5 N10(73)

4 The random variable X has probability density function given by

$$f(x) = \begin{cases} kx(6 - x)^2 & 0 \leqslant x \leqslant 6 \\ 0 & \text{otherwise} \end{cases}$$

where k is a constant.

  (i) Show that $k = \frac{1}{108}$

  (ii) Find $E(X)$.

  (iii) Find the standard deviation of X.

5 The random variable X denotes the number of hours of cloud cover per day at a weather forecasting centre. The probability density function of X is given by

$$f(x) = \begin{cases} \dfrac{(x - 18)^2}{k} & 0 \leqslant x \leqslant 24, \\ 0 & \text{otherwise,} \end{cases}$$

where k is a constant.

  (i) Show that $k = 2016$.

  (ii) On how many days in a year of 365 days can the centre expect to have less than 2 hours of cloud cover?

  (iii) Find the mean number of hours of cloud cover per day.
         Cambridge Paper 7 Q7 J05

6 The continuous random variable X has probability density function given by

$$f(x) = \begin{cases} \frac{3}{4}(x^2 - 1) & 1 \leqslant x \leqslant 2, \\ 0 & \text{otherwise.} \end{cases}$$

  (i) Sketch the probability density function of X.

  (ii) Show that the mean, μ, of X is 1.6875.

  (iii) Show that the standard deviation, σ, of X is 0.2288, correct to 4 decimal places.

  (iv) Find $P(1 \leqslant X \leqslant \mu + \sigma)$.
         Cambridge Paper 7 Q7 J07

7 The time T, in minutes, that Helen has to wait for the bus when she is travelling to work has probability density function given by

$$f(t) = \begin{cases} k & 0 \leqslant t \leqslant 10 \\ 0 & \text{otherwise} \end{cases}$$

  (i) What is the longest time that Helen has to wait for the bus?

  (ii) State the mean time she has to wait for the bus.

  (iii) Find the standard deviation of the time she has to wait for the bus.

  (iv) Find the probability that the time she has to wait is more than 1 standard deviation away from the mean.

8 The continuous random variable X has probability density function given by

$$f(x) = \begin{cases} a + bx & 0 \leqslant x \leqslant 1 \\ 0 & \text{otherwise} \end{cases}$$

(i) Show that $2a + b = 2$.

The median of X is 0.6.

(ii) Find a second equation in a and b and hence find the values of a and b.

9 The time in hours taken for clothes to dry can be modelled by the continuous random variable with probability density function given by

$$f(t) = \begin{cases} k\sqrt{t} & 1 \leqslant t \leqslant 4, \\ 0 & \text{otherwise}, \end{cases}$$

where k is a constant.

(i) Show that $k = \frac{3}{14}$.

(ii) Find the mean time taken for clothes to dry.

(iii) Find the median time taken for clothes to dry.

(iv) Find the probability that the time taken for clothes to dry is between the mean time and the median time.
Cambridge Paper 7 Q7 N08

10 The time in minutes taken by candidates to answer a question in an examination has probability density function given by

$$f(t) = \begin{cases} k(6t - t^2) & 3 \leqslant t \leqslant 6, \\ 0 & \text{otherwise}, \end{cases}$$

where k is a constant.

(i) Show that $k = \frac{1}{18}$.

(ii) Find the mean time.

(iii) Find the probability that a candidate, chosen at random, takes longer than 5 minutes to answer the question.

(iv) Is the upper quartile of the times greater than 5 minutes, equal to 5 minutes or less than 5 minutes? Give a reason for your answer.
Cambridge Paper 7 Q5 J09(71)

11 The average speed of a bus, $x\,\text{km h}^{-1}$, on a certain journey is a continuous random variable X with probability density function given by

$$f(x) = \begin{cases} \dfrac{k}{x^2} & 20 \leqslant x \leqslant 28, \\ 0 & \text{otherwise}. \end{cases}$$

(i) Show that $k = 70$.

(ii) Find E(X).

(iii) Find P($X <$ E(X)).

(iv) Hence determine whether the mean is greater or less than the median.
Cambridge Paper 7 Q6 N02

12 The continuous random variable X has probability density function given by

$$f(x) = \begin{cases} ax^2 + bx & 0 \leqslant x \leqslant 2 \\ 0 & \text{otherwise} \end{cases}$$

where a and b are constants.

The mean of X is 1.25

(i) Show that $b = \frac{3}{4}$ and find the value of a.

(ii) Find the variance of X.

(iii) Verify that the median of X is approximately 1.3

13 The lifetime t, in hours, of a certain type of electrical component can be modelled by a continuous random variable with density function given by

$$f(t) = \begin{cases} 0.05e^{-0.05t} & t > 0 \\ 0 & \text{otherwise} \end{cases}$$

(i) A component is chosen at random from the production line.

Find the probability that:

(a) the component will fail within 12 hours

(b) the component will last longer than 16 hours.

(ii) Find the median lifetime of a component of this type.

(iii) Show that the mean lifetime of a component of this type is 20 hours.

14 The continuous random variable X has probability density function given by

$$f(x) = \begin{cases} k \cos x & 0 \leqslant x \leqslant \frac{1}{2}\pi \\ 0 & \text{otherwise} \end{cases}$$

where k is a constant.

(i) Find the value of k.

(ii) Find E(X).

(iii) Find the median of X.

15 The continuous random variable X has probability density function given by

$$f(x) = \begin{cases} ke^{2x} & 0 \leqslant x \leqslant 4 \\ 0 & \text{otherwise} \end{cases}$$

where k is a constant.

(i) Show that $k = \dfrac{2}{e^8 - 1}$

(ii) Find the mean of X.

16 The continuous random variable X has probability density function given by

$$f(x) = \begin{cases} \dfrac{k}{x^3} & 2 \leqslant x \leqslant 3 \\ 0 & \text{otherwise} \end{cases}$$

where k is a constant.

(i) Find the value of k.

(ii) Show that $E(X) = 2.4$.

(iii) Find the standard deviation of X.

4 Sampling and estimation

In this chapter you will learn about

- samples and populations
- selecting a random sample
- the distribution of the sample mean
- the Central Limit theorem
- unbiased estimates of the population mean and population variance
- confidence intervals for a population mean
- confidence intervals for a population proportion

SAMPLING

Populations

In a statistical enquiry you often need information about a particular group. This group is known as the **population** and it could be small, large or even infinite. Note that the word 'population' does not necessarily mean 'people'.

Here are some examples of populations:

- students in a class
- pebbles on a beach
- people aged 75 or over in a particular city
- cans of soft drink produced in a factory
- fish in Lake Ontario
- rational numbers between 0 and 10.

To collect information about a population you could carry out a census or a sample survey.

Census

In a **census**, every member of the population is surveyed.

When the population is small, this could be straightforward. For example, it would be easy to find out how each student in a class travelled to school on a particular morning. However, when the population is large, taking a census would be very time-consuming and difficult to do with accuracy.

On some occasions it would not be sensible to survey every member. For example, if you performed a census to establish the lifetime of a particular brand of light bulb, you would have to test each bulb until it failed and so you would destroy the population.

Sample survey

As it may be time-consuming, expensive or even impossible to investigate an entire population, it is usual to collect information by selecting a group from the population, known as a **sample**. Provided that the sample is representative of the population, this can give an indication of the population characteristic being studied. Large samples give more reliable information than small samples.

As you want the sample to be **representative** of the population, you should try to **eliminate bias** in the selection of the sample. To do this, select a **random** sample.

Random samples

A **random sample** of size n is a sample chosen such that:

– every member of the population has an equal chance of being selected

– all possible samples of size n have an equal chance of being selected.

If the item being selected is replaced into the population before the next item is selected, then it can appear more than once in the sample. This is known as **sampling with replacement**.

If the item selected is not replaced into the population before the next item is selected, this is known as **sampling without replacement**.

Selecting a random sample

Drawing lots

If you have a small population, then a suitable method may be to 'draw lots'. For example if you wish to select a random sample of 4 students from a class of 20, you could write each name on a piece of paper, fold it and put it into a container. The names would then be drawn at random out of the container, one at a time without replacement until four names had been drawn.

This method, however, is not practical when the population is large.

Using random numbers

A common method of selecting a random sample is to make a list of all the members of the population, then give each member a unique number. For example, if the population is the 590 students in a school, list them in some way (perhaps alphabetically, perhaps by date of birth) and then assign each one a unique number from 001 to 590.

To decide which members of the population to select in the random sample, use random-number tables or generate pseudo-random numbers on a calculator.

Random-number tables

Random-number tables consist of lists containing the digits 0, 1, 2, …, 9 such that each digit has an equal chance of occurring. So, for example, the probability that the digit 3 occurs is $\frac{1}{10}$. The digits may appear singly or be grouped in some way. This is solely for the convenience of printing.

This is an extract from a random-number table where the digits are printed in groups of 4:

  5267   8740   6341   9186   1047   8070   5681   2634   1096   3387   2690

To use this extract to select a random sample of eight students from the 590 students in the school, read off three digits at a time: 526, 787, etc. Ignoring repeats and numbers out of range, continue until you have selected eight three-digit numbers.

Starting at the first number on the left gives the following:

  526   ~~787~~   406   341   ~~918~~   ~~610~~   478   070   568   126   ~~341~~   096
         Out of range   Out of range   Repeat

So you would select the students assigned these numbers:

  526, 406, 341, 478, 070, 568, 126, 096.

It is usual to start at the first number on the left, but you could start anywhere in the tables and go forwards or backwards.

Example 4.1

The editor of a magazine wishes to obtain the views of a random sample of readers about the future of the magazine.

(i) A sub-editor proposes that they include in one issue of the magazine a questionnaire for readers to complete and return. Give two reasons why the readers who return the questionnaire would not form a random sample.

The editor decides to use a table of random numbers to select a random sample of 50 readers from the 7302 regular readers. These regular readers are numbered from 1 to 7302. The first few random numbers that the editor obtains from the table are as follows:

49757   80239   52038   60882

(ii) Use these random numbers to select the first three members in the sample.

Cambridge Paper 7 Q2 N10(73)

(i) Only readers of that particular issue of the magazine would see the questionnaire and have the opportunity to respond.

Only the more committed readers may respond, introducing bias.

(ii) The regular readers are numbered 0001 to 7302.

Read off four digits at a time, disregarding any repeats or numbers out of range:

4975   7802   3952   0386   Ignore 7802 as it is out of range.

So, you could select the readers assigned the numbers 4975, 3952, 0386.

Calculator random-number generator

You probably have a **random-number generator** key on your calculator which produces a number every time you press it. The numbers generated are in fact obtained using a mathematical formula and are *pseudo*-random numbers. However, they suit the purpose very well indeed.

To generate a pseudo-random number on your calculator, press [Ran] [=]. Then each time you press [=], a number will be generated.

The number generated usually appears as zero followed by a decimal point, then three digits, for example

0.507   0.673   0.489   0.233   0.362   0.002

Use the numbers after the decimal point to give the random digits

507   673   489   233   362   002

If there are fewer than three digits after the decimal point, fill in with zeros, to the right, until you have 3 digits. For example, 0.26 gives the digits 260 and 0.3 gives the digits 300.

Exercise 4a

1 (i) Explain briefly the difference between a census and a sample survey. Give an example to illustrate the practical use of each method.

A school held an evening disco which was attended by 500 students. The disco organisers were keen to assess the success of the evening. Having decided to obtain information from those attending the disco, they were undecided as to whether to use a census or a sample survey.

(ii) Which method would you recommend they use? Give one advantage and one disadvantage associated with your recommendation.

2 Use the following extract from a table of random digits to select a random sample of 8 people from a group of 50 people. Explain your method clearly.

| 02596 | 77442 | 38703 | 85614 |
| 12634 | 19186 | 10478 | 96200 |

3 A union wishes to carry out a survey to gather the opinions of employees at a factory on new proposals on working hours. The union plans to hand out a questionnaire to the first 50 employees arriving at the next union meeting.

(i) Give a reason why this method will give an unsatisfactory sample for the purposes of the survey.

(ii) Describe briefly a satisfactory method for conducting the survey.

4 Sybil is doing a survey of adults in her town about a proposed bypass. She plans to choose 25 residential numbers at random from the telephone directory for her town and question an adult from the house.

(i) Give two reasons why this method will not produce a random sample.

(ii) Describe a method that will produce a random sample.

5 Johann plans to use the random number generator on his calculator to select a random sample of 10 students from the 70 students in his year at college.

(i) Describe briefly what he needs to do before using his calculator.

(ii) Johann obtains the following random numbers from his calculator.

| 234 | 568 | 892 | 303 | 710 | 943 |
| 629 | 342 | 153 | 376 | 184 | |

Write down the 10 numbers he could use to select the sample.

SAMPLE STATISTICS: THE SAMPLE MEAN $\overline{X}$

When you are trying to find out information about a population it seems sensible to take random samples and then consider the values obtained from them. It is therefore useful to know how these sample values are distributed.

Imagine carrying out the following procedure:

- Take a random sample of n independent observations from a population.

- Calculate the mean of these n sample values. This is known as the **sample mean**.

- Repeat the procedure until you have taken all possible samples of size n, calculating the mean of each sample.

- Form a distribution of all the sample means.

The distribution that would be formed is called the **distribution of the sample mean**.

The mean and variance of the distribution of the sample mean

It is possible to work out the mean and variance of the distribution of the sample mean using the results in chapter 1.

Consider the random variable X, where $E(X) = \mu$ and $Var(X) = \sigma^2$.

Take n independent observations $X_1, X_2, ..., X_n$ of X.

Since $E(X) = \mu,$

$E(X_1) = \mu, \ E(X_2) = \mu, \ ... \ , \ E(X_n) = \mu$

Since $Var(X) = \sigma^2$

$Var(X_1) = \sigma^2, \quad Var(X_2) = \sigma^2, \ ... \ , \quad Var(X_n) = \sigma^2$

The sample mean is

$$\overline{X} = \frac{X_1 + X_2 + \ldots + X_n}{n}$$

$$= \frac{1}{n}X_1 + \frac{1}{n}X_2 + \ldots + \frac{1}{n}X_n$$

$$\mathrm{E}(\overline{X}) = \mathrm{E}\left(\frac{1}{n}X_1 + \frac{1}{n}X_2 + \ldots + \frac{1}{n}X_n\right)$$

$$= \mathrm{E}\left(\frac{1}{n}X_1\right) + \mathrm{E}\left(\frac{1}{n}X_2\right) + \ldots + \mathrm{E}\left(\frac{1}{n}X_n\right)$$

$$= \frac{1}{n}\mathrm{E}(X_1) + \frac{1}{n}\mathrm{E}(X_2) + \ldots + \frac{1}{n}\mathrm{E}(X_n) \qquad \text{Using } \mathrm{E}(aX) = a\mathrm{E}(X) \text{ page 26}$$

$$= \frac{1}{n}\mu + \frac{1}{n}\mu + \ldots + \frac{1}{n}\mu$$

$$= n \times \frac{1}{n}\mu$$

$$= \mu$$

$$\mathrm{Var}(\overline{X}) = \mathrm{Var}\left(\frac{1}{n}X_1 + \frac{1}{n}X_2 + \ldots + \frac{1}{n}X_n\right)$$

$$= \mathrm{Var}\left(\frac{1}{n}X_1\right) + \mathrm{Var}\left(\frac{1}{n}X_2\right) + \ldots + \mathrm{Var}\left(\frac{1}{n}X_n\right)$$

$$= \left(\frac{1}{n}\right)^2 \mathrm{Var}(X_1) + \left(\frac{1}{n}\right)^2 \mathrm{Var}(X_2) + \ldots + \left(\frac{1}{n}\right)^2 \mathrm{Var}(X_n) \qquad \text{Using } \mathrm{Var}(aX) = a^2\mathrm{Var}(X) \text{ page 26}$$

$$= \frac{1}{n^2}\sigma^2 + \frac{1}{n^2}\sigma^2 + \ldots + \frac{1}{n^2}\sigma^2$$

$$= n \times \frac{1}{n^2}\sigma^2$$

$$= \frac{\sigma^2}{n}$$

So, $\mathrm{E}(\overline{X}) = \mathrm{E}(X) = \mu$

$$\mathrm{Var}(\overline{X}) = \frac{\mathrm{Var}(X)}{n} = \frac{\sigma^2}{n}$$

You should learn these results but you do not need to learn the proofs

The standard deviation of the distribution of sample means is $\sqrt{\dfrac{\sigma^2}{n}}$, usually written as $\dfrac{\sigma}{\sqrt{n}}$. This is known as the **standard error** of the mean. Since σ has been divided by $\sqrt{n}$, the standard deviation of $\overline{X}$ is smaller than the standard deviation of X, indicating that the sample means are much more clustered around μ than the population values.

The distribution of $\overline{X}$ when X is normally distributed

If $X \sim \mathrm{N}(\mu, \sigma^2)$ and $X_1, X_2, \ldots, X_n$ are n independent observations taken from this **normal** distribution, then the distribution of $\overline{X}$ is **also normal**, where $\overline{X} = \dfrac{X_1 + X_2 + \ldots + X_n}{n}$.

So, if $X \sim \mathrm{N}(\mu, \sigma^2)$, then $\overline{X} \sim \mathrm{N}\left(\mu, \dfrac{\sigma^2}{n}\right)$.

This result holds true even for small values of n.

The following diagrams help to illustrate the shape of the distribution of the sample means resulting from different sized samples taken from a normal distribution with mean 100 and standard deviation 8.

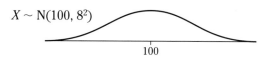
$X \sim N(100, 8^2)$

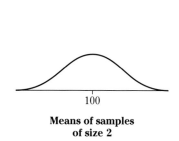

Means of samples of size 2

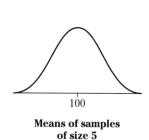

Means of samples of size 5

Means of samples of size 25

You can see from the diagrams that the sample means are more clustered around μ than the population values are. In fact, the larger the sample size, the more clustered they become.

Example 4.2

The lengths of time people take to complete a certain type of puzzle are normally distributed with mean 48.8 minutes and standard deviation 15.6 minutes. The random variable X represents the time taken in minutes by a randomly chosen person to solve this type of puzzle. The times taken by random samples of 5 people are noted. The mean time $\overline{X}$ is calculated for each sample.

(i)   State the distribution of $\overline{X}$, giving the values of any parameters.

(ii) Find $P(\overline{X} < 50)$.

Cambridge Paper 7 Q2 J08

X is the time, in minutes, taken to solve the puzzle, where

$$X \sim N(48.8, 15.6^2) \qquad \mu = 48.8, \sigma = 15.6$$

(i) $\overline{X}$ is the mean time, in minutes, taken by 5 people to solve the puzzle.

$$E(\overline{X}) = 48.8, \qquad Var(\overline{X}) = \frac{15.6^2}{5}$$

Since the distribution of X is normal, the distribution of $\overline{X}$ is also normal.

So $\overline{X} \sim N\left(48.8, \dfrac{15.6^2}{5}\right) \qquad \text{s.d.} = \dfrac{15.6}{\sqrt{5}}$

To avoid rounding errors, do not calculate the s.d. at this stage but leave it as $\dfrac{15.6}{\sqrt{5}}$

(ii) *Standardise using*

$$Z = \frac{\overline{X} - \mu}{\frac{\sigma}{\sqrt{n}}}$$

$$P(\overline{X} < 50) = P\left(Z < \frac{50 - 48.8}{\frac{15.6}{\sqrt{5}}}\right)$$

$$= P(Z < 0.172)$$

$$= \Phi(0.172)$$

$$= 0.5683$$

$$= 0.568 \text{ (3 s.f.)}$$

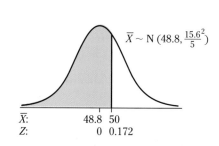
$\overline{X} \sim N(48.8, \frac{15.6^2}{5})$

$\overline{X}$:   48.8 50
Z: 0   0.172

Example 4.3

The heights of a new variety of sunflower can be modelled by a normal distribution with mean 200 cm and standard deviation 40 cm.

(i) A random sample of 50 sunflowers is taken and the mean height calculated. Find the probability that the sample mean lies between 195 cm and 205 cm.

(ii) A total of 100 samples, each of 50 observations, are taken. In how many of these would you expect the sample mean to lie between 195 cm and 205 cm?

Let X be the height, in centimetres, of a sunflower, where

$$X \sim N(200, 40^2) \qquad\qquad \mu = 200, \sigma = 40$$

(i) $\bar{X}$ is the mean height, in cm, of 50 sunflowers.

$$E(\bar{X}) = 200, \qquad Var(\bar{X}) = \frac{40^2}{50}$$

So $\quad \bar{X} \sim N\left(200, \frac{40^2}{50}\right) \qquad$ s.d. $= \frac{40}{\sqrt{50}}$

$$P(195 < \bar{X} < 205) = P\left(\frac{195 - 200}{\frac{40}{\sqrt{50}}} < Z < \frac{205 - 200}{\frac{40}{\sqrt{50}}}\right)$$

$$P(-a < Z < a)$$
$$= \Phi(a) - (1 - \Phi(a))$$
$$= \Phi(a) - 1 + \Phi(a)$$
$$= 2\Phi(a) - 1$$

$$= P(-0.884 < Z < 0.884)$$

$$= 2\Phi(0.884) - 1$$

$$= 2 \times 0.8117 - 1$$

$$= 0.6234$$

$$= 0.623 \text{ (3 s.f.)}$$

$\bar{X} \sim N\left(200, \frac{40^2}{50}\right)$

$\bar{X}$:	195	200	205
Z:	−0.884	0	0.884

(ii) $100 \times 0.6234 = 62.34$

So you would expect approximately 62 of the samples to have a mean between 195 cm and 205 cm.

Example 4.4

At a college, the weights of male students may be modelled by a normal distribution with mean 70 kg and standard deviation 5 kg.

(i) Find the probability that the mean height of a random sample of 4 male students is less than 65 kg.

(ii) The probability that the mean weight of a random sample of n students is less than 73 kg is 0.9918

Find the value of n.

X is the weight, in kilograms, of a male student at the college, where

$$X \sim N(70, 5^2) \qquad\qquad \mu = 70, \sigma = 5$$

(i) $\bar{X}$ is the mean weight, in kilograms, of 4 students at the college.

$$E(\bar{X}) = 70, \qquad \text{Var}(\bar{X}) = \frac{5^2}{4}$$

Since the distribution of X is normal, the distribution of $\bar{X}$ is also normal.

So $\bar{X} \sim N\left(70, \dfrac{5^2}{4}\right)$ s.d. $= \dfrac{5}{\sqrt{4}}$

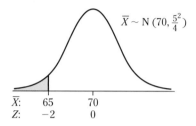

$$\begin{aligned}
P(\bar{X} < 65) &= P\left(Z < \frac{65-70}{\frac{5}{\sqrt{4}}}\right) \\
&= P(Z < -2) \\
&= 1 - \Phi(2) \\
&= 0.0228
\end{aligned}$$

Note: The diagram below shows the distributions of X and $\bar{X}$, drawn to the same scale.

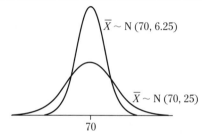

(ii) $\bar{X}$ is the mean weight, in kilograms, of n students at the college.

$$\bar{X} \sim N\left(70, \frac{5^2}{n}\right) \qquad \text{s.d.} = \frac{5}{\sqrt{n}}$$

$$\begin{aligned}
P(\bar{X} < 73) &= P\left(Z < \frac{73-70}{\frac{5}{\sqrt{n}}}\right) \\
&= P\left(Z < \frac{3\sqrt{n}}{5}\right)
\end{aligned}$$

Note that

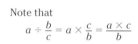

$$a \div \frac{b}{c} = a \times \frac{c}{b} = \frac{a \times c}{b}$$

But you are given that $P(\bar{X} < 73) = 0.9918$

so $P\left(Z < \dfrac{3\sqrt{n}}{5}\right) = 0.9918$

Using the normal table in reverse,

$$\Phi^{-1}(0.9918) = 2.40$$

so $\dfrac{3\sqrt{n}}{5} = 2.40$

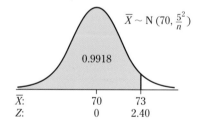

$$\sqrt{n} = \frac{5 \times 2.40}{3} = 4$$

$$n = 16$$

Exercise 4b

1   The volume of wine in bottles is normally distributed with a mean 758 ml and a standard deviation 12 ml.

A random sample of 10 bottles is taken and the mean volume found. Find the probability that the sample mean is:

(i) less than 750 ml

(ii) between 752 ml and 760 ml.

2   A wholesaler sells carrots whose masses are normally distributed with mean 145 g and standard deviation 9 g.

A sample of 16 carrots is taken. Stating a necessary assumption, find the probability that the mean mass of a carrot in the sample is:

(i) less than 148 g

(ii) greater than 141 g

(iii) between 143 g and 150 g.

3 In a certain city, men have heights distributed normally with mean 170 cm and standard deviation 10 cm.

(i) Find the probability that the mean height of 3 randomly selected men is greater than 178 cm.

(ii) Find the probability that the mean height of a random sample of 20 men differs from the population mean height by more than 10 cm.

4 The length of a particular type of insect may be modelled by a normal distribution with mean 15.3 mm and standard deviation 2.6 mm.

For a biology project, each student in a class of 30 students collects a random sample of 10 insects of this type and finds the mean length.

(i) Jenny is a student in the class. Find the probability that the mean length of the insects in Jenny's sample is between 13.5 mm and 15.1 mm.

(ii) How many of the 30 samples collected by the students in the class would you expect to have a mean length greater than 16.0 mm?

5 The random variable X has a distribution N(30, 5).

Random samples of size n are taken from the distribution.

(i) Given that $n = 10$, find the probability that the mean of the sample exceeds 30.5

(ii) Given that $n = 40$, find the probability that the mean of the sample exceeds 30.5

(iii) Given that $n = 100$, find the probability that the mean of the sample exceeds 30.5

(iv) Find the least value of n such that the probability that the sample mean exceeds 30.5 is less than 1%.

6 The random variable Y follows a normal distribution with mean 25 and variance 340. A random sample of size n is taken from the distribution. Find the least value of n such that $P(\bar{Y} > 28) < 0.005$

7 In a computer game, competitors have to find their way out of a maze. The time taken, in seconds, may be modelled by a normal distribution with mean 80 and standard deviation σ.

The probability that the mean time of a random sample of 100 competitors is greater than 82 seconds is 0.05

Find the value of σ.

8 The random variable X is distributed normally with mean 82.6 and standard deviation σ. The mean of 20 independent observations of X is $\bar{X}$. Given that $P(\bar{X} > 80) = 0.75$, find the value of σ.

The distribution of $\bar{X}$ when X is not normally distributed

The following diagrams illustrate the distribution of X, together with the distribution of $\bar{X}$, based on 50 random samples of size 10, 15 and 30 taken from X, where X follows a binomial distribution, a Poisson distribution and a continuous uniform distribution.

(i) $X \sim B(10, 0.25)$

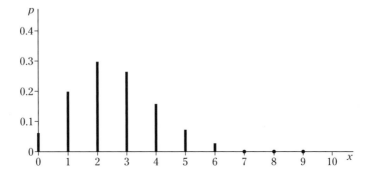

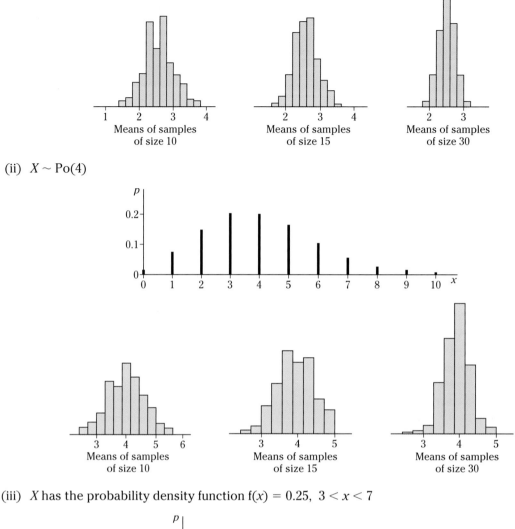

(ii) $X \sim \text{Po}(4)$

(iii) X has the probability density function $f(x) = 0.25, \; 3 < x < 7$

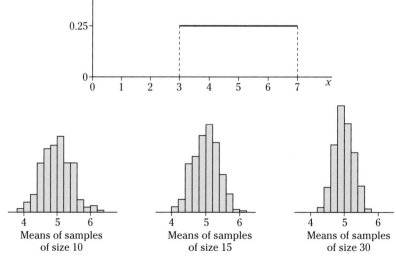

From the diagrams you can see that when samples are taken from a population that is **not normally distributed**, the distribution of $\overline{X}$ takes on the characteristic normal shape as the sample size increases. This is a surprising result. It leads to a very important theorem, called the Central Limit theorem.

The Central Limit theorem

The **Central Limit theorem** (CLT) states that for samples of size n taken from a **non-normal population** with mean μ and variance σ^2, the distribution of $\overline{X}$ tends to a normal distribution as n tends to infinity. This means that, when **n is large** ($n \geqslant 30$ say), the distribution of $\overline{X}$ is **approximately normal**.

If X has mean μ and variance σ^2, then, **for large n**, ($n \geqslant 30$), by the Central Limit theorem,

$$\overline{X} \sim N\left(\mu, \frac{\sigma^2}{n}\right) \text{ approximately.}$$

The larger the sample size n, the better the approximation.

You now have two very important results:

1 If $X \sim N(\mu, \sigma^2)$, then $\overline{X} \sim N\left(\mu, \frac{\sigma^2}{n}\right)$, even for small values of n.

2 If X is not normally distributed but has mean μ and variance σ^2, then $\overline{X} \sim N\left(\mu, \frac{\sigma^2}{n}\right)$, approximately, provided n is sufficiently large.

Example 4.5

Over a long period of time it is found that the amount of sunshine on any day in a particular town in Spain has mean 6.7 hours and standard deviation 3.1 hours.

(i) Find the probability that the mean amount of sunshine over a random sample of 300 days is between 6.5 and 6.8 hours.

(ii) Give a reason why it is not necessary to assume that the daily amount of sunshine is normally distributed in order to carry out the calculation in part (i).

Cambridge Paper 7 Q2 N04

Let X be the daily amount of sunshine, in hours, where

$$X \sim N(6.7, 3.1^2) \qquad\qquad \mu = 6.7, \ \sigma = 3.1$$

(i) $\overline{X}$ is the mean amount of sunshine, in hours, over a random sample of 300 days.

$$E(\overline{X}) = 6.7, \qquad Var(\overline{X}) = \frac{3.1^2}{300}$$

Since n is large, by the Central Limit theorem,

$$\overline{X} \sim N\left(6.7, \frac{3.1^2}{300}\right) \text{ approximately} \qquad \text{s.d.} = \frac{3.1}{\sqrt{300}}$$

$$P(6.5 < \overline{X} < 6.8) = P\left(\frac{6.5 - 6.7}{\dfrac{3.1}{\sqrt{300}}} < Z < \frac{6.8 - 6.7}{\dfrac{3.1}{\sqrt{300}}}\right) \qquad P(-a < Z < b) = \Phi(b) - (1 - \Phi(a))$$

$$= P(-1.117 < Z < 0.559)$$

$$= \Phi(0.559) - (1 - \Phi(1.117))$$

$$= 0.7119 - (1 - 0.8679)$$

$$= 0.5798$$

$$= 0.580 \text{ (3 s.f.)}$$

$\overline{X} \sim N\left(6.7, \frac{3.1^2}{300}\right)$

| $\overline{X}$: | 6.5 | 6.7 | 6.8 |
| Z: | -1.117 | 0 | 0.559 |

(ii) It is not necessary to assume that the daily amount of sunshine is normally distributed because the sample size n is large ($n = 300$) and the Central Limit theorem states that, for large n, the distribution of $\overline{X}$ is approximately normal.

Special note when *X* is a discrete variable

You will recall that when the normal distribution is used as an approximation to a discrete distribution, such as a binomial distribution (S1 chapter 6) or a Poisson distribution (S2 chapter 1), a continuity correction of $\pm \frac{1}{2}$ is needed.

In a similar way, when the normal approximation is applied to the distribution of $\overline{X}$, the mean of a random sample taken from a **discrete distribution**, a **continuity correction** of $\pm \frac{1}{2n}$ should be used, where *n* is the number of observations in the sample. However, as *n* tends to infinity, $\frac{1}{2n}$ tends to zero. So, for large sample sizes, the continuity correction makes little difference to the answer. It is therefore often omitted in questions relating to the distribution of $\overline{X}$.

You will not need to use a continuity correction when considering the distribution of $\overline{X}$ in the examination.

Example 4.6

The random variable *X* has the distribution B(12, 0.4). Find the probability that the mean of a random sample of 100 independent observations of *X* is greater than 5.

$X \sim \text{B}(12, 0.4)$. *X* follows a **binomial** distribution, which is discrete.

$\mu = 12 \times 0.4 = 4.8$ Recall from S1: When $X \sim \text{B}(n, p)$, mean $= np$, variance $= npq$ where $q = 1 - p$

$\sigma^2 = 12 \times 0.4 \times 0.6 = 2.88$

$\overline{X}$ is the mean of a sample of 100 observations, so $n = 100$.

$$\text{E}(\overline{X}) = 4.8, \qquad \text{Var}(\overline{X}) = \frac{2.88}{100}$$

Since *n* is large, by the Central Limit theorem,

$\overline{X} \sim \text{N}\left(4.8, \frac{2.88}{100}\right)$ approximately   s.d. $= \sqrt{\frac{2.88}{100}} = \sqrt{0.0288}$

$$P(\overline{X} > 5) = P\left(Z > \frac{5 - 4.8}{\sqrt{0.0288}}\right)$$

$$= P(Z > 1.179)$$

$$= 1 - \Phi(1.179)$$

$$= 1 - 0.8808$$

$$= 0.1192$$

$$= 0.119 \text{ (3 sf)}$$

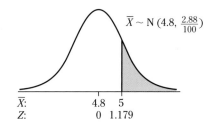

Example 4.7

The number of telephone calls made in an evening to a counselling service may be modelled by a Poisson distribution with parameter 6.2. A random sample of 45 evenings is taken. Find the probability that the mean number of calls per evening in the sample is less than 5.1.

Let X be the number of telephone calls in an evening, where $X \sim \text{Po}(6.2)$.

X is known to follow a **Poisson** distribution, which is discrete.

$$\mu = 6.2, \qquad \sigma^2 = 6.2$$

Recall from chapter 1: When $X \sim \text{Po}(\lambda)$, mean = λ, variance = λ.

$\overline{X}$ is the mean number of telephone calls per evening in a random sample of 45 evenings, so $n = 45$.

$$\text{E}(\overline{X}) = 6.2, \qquad \text{Var}(\overline{X}) = \frac{6.2}{45}$$

Since n is large, by the Central Limit theorem,

$$\overline{X} \sim \text{N}\left(6.2, \frac{6.2}{45}\right) \text{ approximately} \qquad \text{s.d.} = \sqrt{\frac{6.2}{45}} = \sqrt{0.1377\ldots}$$

$$P(\overline{X} < 5.1) = P\left(Z < \frac{5.1 - 6.2}{\sqrt{0.1377}}\right)$$

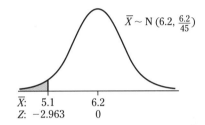

$\overline{X} \sim \text{N}\left(6.2, \frac{6.2}{45}\right)$

$$= P(Z < -2.963)$$

$$= 1 - \Phi(2.963)$$

$$= 1 - 0.9985$$

$$= 0.0015$$

| $\overline{X}$: | 5.1 | 6.2 |
| Z: | −2.963 | 0 |

Example 4.8

Sid arrives at random times at a bus stop. The time X, in minutes, that he has to wait for the next bus is given by

$$f(x) = \begin{cases} \frac{1}{9} & 0 \leqslant x \leqslant 9 \\ 0 & \text{otherwise} \end{cases}$$

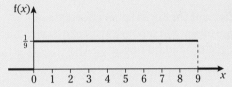

(i)   Find the mean and variance of X.

(ii) Find the probability that the mean time he has to wait on a random sample of 75 occasions is less than 5 min.

(i) X is the waiting time, in minutes, where

$$f(x) = \frac{1}{9} \text{ for } 0 \leqslant x \leqslant 9$$

This is a continuous distribution, known as the uniform distribution (see Chapter 3, page 62)

By symmetry, $\mu = \text{E}(X) = 4.5$

$$\text{Var}(X) = \int_a^b x^2 \, f(x) \, dx - \mu^2$$

See page 79

$$= \int_0^9 \frac{1}{9} x^2 \, dx - 4.5^2$$

$$= \frac{1}{9}\left[\frac{x^3}{3}\right]_0^9 - 20.25$$

$$= \frac{1}{9}\left(\frac{9^3}{3} - 0\right) - 20.25$$

$$= 6.75$$

(ii) $\overline{X}$ is the mean waiting time over a random sample of 75 observations, so $n = 75$.

$$E(\overline{X}) = 4.5, \qquad Var(\overline{X}) = \frac{6.75}{75}$$

Since n is large, by the Central Limit theorem,

$$\overline{X} \sim N\left(4.5, \frac{6.75}{75}\right) \text{ approximately} \qquad \text{s.d.} = \sqrt{\frac{6.75}{75}} = 0.3$$

$$P(\overline{X} < 5) = P\left(Z < \frac{5 - 4.5}{0.3}\right)$$

$$= P(Z < 1.667)$$

$$= \Phi(1.667)$$

$$= 0.9522$$

$$= 0.952 \text{ (3 s.f.)}$$

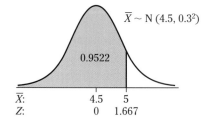

$\overline{X} \sim N(4.5, 0.3^2)$

0.9522

$\overline{X}$:   4.5   5

Z: 0   1.667

Exercise 4c

1 In an examination taken by students in a particular country, the mean mark was 64.5 and the variance was 64.

A total of 100 examination scripts are selected at random. Find the probability that the mean mark of the scripts in the sample is:

(i) higher than 65.5

(ii) between 63.8 and 64.5

2 A manufacturer produces small chocolate cakes. The amount of mixture in a cake has mean 52 g and standard deviation 4 g.

(i) Find the probability that the mean amount of mixture in a random sample of 75 cakes is less than 51 g.

(ii) Give a reason why it is not necessary to assume that the amount of mixture in a cake is normally distributed in order to carry out the calculations in part (i).

3 Rubber balls are produced for sale in toy shops. The diameters X, in millimetres, of the balls have mean 62.5 and standard deviation 1.2.

The balls are packed in boxes of 50.

Given that the balls in a box may be regarded as a random sample, find the probability that the mean diameter of the balls in a box is:

(i) less than 62.6 mm

(ii) between 62.0 mm and 62.3 mm.

4 The random variable X has the distribution B(12, 0.4).

(i) Find the mean and variance of X.

A random sample of 50 observations of X is taken.

(ii) Describe fully the distribution of the sample mean.

(iii) Find the probability that the sample mean is:

(a) less than 4.7   (b) more than 5.4

5 The number of text messages Joe sends to his girlfriend each day may be modelled by a Poisson distribution with mean 12.5

Find the probability that, in a random sample of 40 days, the mean number of text messages Joe sends is greater than 14.

6 The continuous random variable X has a probability density function given by

$$f(x) = \begin{cases} \frac{3}{8}(1 + x^2) & -1 \leqslant x \leqslant 1 \\ 0 & \text{otherwise} \end{cases}$$

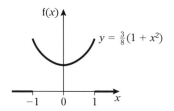

$y = \frac{3}{8}(1 + x^2)$

(i) **State** the value of $E(X)$.

(ii) Find $Var(X)$.

The mean of 50 independent observations of X is denoted by $\overline{X}$.

(iii) Find the mean and variance of $\overline{X}$.

(iv) State the approximate distribution of $\overline{X}$.

(v) Find $P(0 < \overline{X} < 0.2)$

7 The mass, in kilograms, of articles produced by a machine has mean μ and standard deviation 4.55

A random sample of 100 articles from the machine is taken and the mean mass calculated.

(i) Find the probability that the sample mean will differ from μ by less than 0.8 kg.

(ii) State, with a reason, whether it was necessary to use the Central Limit theorem in part (i).

8 Random samples of size 65 are taken from the distribution Po(5.8).

(i) Describe fully the distribution of the sample mean.

(ii) Find the probability that the sample mean lies between 5.1 and 6.3

UNBIASED ESTIMATES

To define

- a binomial distribution, you need to know n and p
- a Poisson distribution, you need to know λ
- a normal distribution, you need to know μ and σ^2.

These are known as the **population parameters** of the distributions.

If you do not know the value of a particular population parameter, you could take a random sample from the distribution and use it in some way to make an **estimate** of the value of your unknown parameter.

This estimate is **unbiased** if the mean of a large number of values taken in the same way is equal to the true value of the parameter.

Unbiased estimates of population mean and variance

Taking a random sample of size n:

An unbiased estimate of the population mean is $\hat{\mu}$, where

$$\hat{\mu} = \overline{x} = \frac{\Sigma x}{n}$$

An unbiased estimate of the population variance is s^2, where

$$s^2 = \frac{1}{n-1}\left(\Sigma x^2 - \frac{(\Sigma x)^2}{n}\right)$$

The notation $\tilde{\sigma}^2$ is sometimes used instead of s^2.

The above formulae are given on the List of Formulae available in the examination.

Alternatively, $s^2 = \dfrac{n}{n-1} \times$ sample variance.

Note that, since $\dfrac{n}{n-1} > 1$, the unbiased estimate of the population variance is greater than the sample variance.

The following working shows that the two formulae are equivalent:

$$s^2 = \frac{n}{n-1} \times \text{sample variance}$$

$$= \frac{n}{n-1}\left(\frac{\Sigma x^2}{n} - \left(\frac{\Sigma x}{n}\right)^2\right)$$

$$= \frac{n}{n-1}\left(\frac{\Sigma x^2}{n}\right) - \frac{n}{n-1}\left(\frac{\Sigma x}{n}\right)\left(\frac{\Sigma x}{n}\right)$$

$$= \frac{1}{n-1}\left(\Sigma x^2 - \frac{(\Sigma x)^2}{n}\right)$$

This proof is not needed for the examination.

Example 4.9

A magazine conducted a survey about the sleeping time of adults. A random sample of 12 adults was chosen from the adults travelling to work on a train.

  (i) Give a reason why this is an unsatisfactory sample for the purpose of the survey.

 (ii) State a population for which this sample would be satisfactory.

     A satisfactory sample of 12 adults gave numbers of hours of sleep as shown below.

       4.6   6.8   5.2   6.2   5.7   7.1   6.3   5.6   7.0   5.8   6.5   7.2

(iii) Calculate unbiased estimates of the mean and variance of the sleeping times of adults.

                                                              Cambridge Paper 7 Q1 J08

  (i) The adults travelling to work on the train are not representative of all adults.

 (ii) A satisfactory population would be 'adults who travel to work on this train'.

(iii) Using a calculator in SD mode:

$$\Sigma x = 74, \qquad \Sigma x^2 = 463.56$$

$$\hat{\mu} = \bar{x} = \frac{\Sigma x}{n} = \frac{74}{12} = 6.166\ldots = 6.17 \text{ (3 s.f.)}$$

$$s^2 = \frac{1}{n-1}\left(\Sigma x^2 - \frac{(\Sigma x)^2}{n}\right)$$

$$= \frac{1}{11}\left(463.56 - \frac{74^2}{12}\right)$$

$$= 0.6569\ldots$$

$$= 0.657 \text{ (3 s.f.)}$$

Calculator notes

- If you are given raw data you can get the values of $\bar{x}$ and s directly from your calculator in **statistical mode**. The unbiased estimate, s, of the population standard deviation is given by the key usually marked $\boxed{x\sigma_{n-1}}$. On some calculators it is a function available in STAT mode. If you are asked for the unbiased estimate of the variance, remember to square the value of s.

- It is a good idea to show the values of Σx and Σx^2 substituted into the formulae. If you do not show any working and your answer is wrong, you will not be awarded any marks.

- Double-check all calculations!

Data is sometimes given only in summarised form. In this case the formulae must be used.

Example 4.10

The heights x, in centimetres, of a random sample of 250 men undergoing a particular medical examination were measured. The data are summarised by $\Sigma x = 43\,205$ and $\Sigma x^2 = 7\,469\,107$.

Calculate unbiased estimates of the population mean and standard deviation.

$$\hat{\mu} = \bar{x} = \frac{\Sigma x}{n} = \frac{43205}{250} = 172.82$$

$$s^2 = \frac{1}{n-1}\left(\Sigma x^2 - \frac{(\Sigma x)^2}{n}\right)$$

$$= \frac{1}{249}\left(7469107 - \frac{43205^2}{250}\right)$$

$$= 9.7144...$$

$$s = \sqrt{9.7144...}$$

$$= 3.116...$$

$$= 3.12 \text{ (3 s.f.)}$$

Exercise 4d

1 The concentrations, in milligrams per litre, of a trace element in 7 randomly chosen samples of water from a spring were as follows:

240.8 237.3 236.7 236.6 234.2 233.9 232.5

Find unbiased estimates of the mean and the variance of the concentration of the trace element per litre of water from the spring.

2 Find unbiased estimates of the mean μ and variance σ^2 of the population from which each of the following samples is taken:

(i) 46, 48, 51, 50, 45, 53, 50, 48

(ii) 1.684, 1.691, 1.687, 1.688, 1.689, 1.688, 1.690, 1.693, 1.685

(iii)

x	20	21	22	23	24	25
f	4	14	17	26	20	9

(iv)

Mass in grams	$10 \leqslant x < 20$	$20 \leqslant x < 30$	$30 \leqslant x < 40$	$40 \leqslant x < 50$
Frequency	2	12	27	9

3 Find unbiased estimates of the mean μ and variance σ^2 of the population from which each of the following samples is taken:

(i) $\Sigma x = 120$, $\Sigma x^2 = 2102$, $n = 8$

(ii) $\Sigma x = 100$, $\Sigma x^2 = 1028$, $n = 10$

(iii) $\Sigma x = 330$, $\Sigma x^2 = 23\,700$, $n = 34$

4 A measuring rule was used to measure the length of a rod of stated length 1 metre. On 8 successive occasions the following results, in millimetres, were obtained:

1000 999 999 1002 1001 1000 1002 1001

Calculate unbiased estimates of the population mean and the standard deviation of the errors, in millimetres, occurring when the rule is used for measuring a 1 m length.

5 Cartons of orange juice are filled by a machine. A sample of 10 cartons selected at random from the production contained the following volumes of orange juice, in millilitres:

201.2 205.0 209.1 202.3 204.6
206.4 210.1 201.9 203.7 207.3

(i) Calculate unbiased estimates of the population mean and variance of the volume of orange juice dispensed into the carton by the machine.

(ii) Is the unbiased estimate of the population variance greater than, equal to or less than the sample variance?

6 A random sample of size 26 is taken from a population. The variance of the sample is 5.34

Calculate the unbiased estimate of the population standard deviation.

7 A random sample of 100 plants of a certain type are chosen and their heights, x cm, measured. The results are summarised by

$$\Sigma x = 7600, \qquad \Sigma x^2 = 592\,000$$

Calculate unbiased estimates of the population mean and variance.

8 The weights of bags of flour of a particular brand are normally distributed. A random sample of 150 bags were weighed and the weights, x grams, are summarised by

$$\Sigma x = 112\,200, \qquad \Sigma x^2 = 83\,927\,544$$

(i) Calculate unbiased estimates of the population mean and variance.

(ii) Using the values found in part (i), find the probability that the mean weight of a random sample of size 50 will be greater than 749 grams.

9 80 adults in Ruritania were asked to measure their pulse rates, x beats per minute, when they woke up in the morning. The results were collected together and summarised as follows.

$$\Sigma x = 5520, \qquad \Sigma x^2 = 382\,160$$

Stating an assumption about the 80 adults, calculate unbiased estimates of the mean and variance of the pulse rates, upon waking, of all adults in Ruritania.

CONFIDENCE INTERVALS FOR μ, THE POPULATION MEAN

Another way of using a sample value to give a good idea of an unknown population parameter is to construct an **interval**, known as a **confidence interval** (sometimes abbreviated 'C. I.').

This is an interval that has a specified probability of **including** the parameter. The confidence interval is usually written (a, b) and the end-values, a and b, are known as **confidence limits**. The probabilities most often used in confidence intervals are 90%, 95% and 99%

Suppose you do not know the population mean, μ, and you want to work out a **95% confidence interval** for it. You would need to construct an interval (a, b) so that

$$P(a < \mu < b) = 0.95$$

i.e. the probability that the interval includes μ (or has 'trapped' μ) is 0.95 or 95%

The interval that you construct uses the value of $\overline{x}$, the mean of a random sample of size n taken from the population.

The following computer simulation illustrates the intervals obtained when 100 confidence intervals were constructed, each with 95% confidence. The dotted line shows the true value of μ.

In this simulation six out of the 100 (shown in bold) do not include μ.

However, **on average**, 95% of the confidence intervals constructed in this way include the population mean. So, **on average**, 5% of the confidence intervals do not include μ.

In practice, you only construct *one* interval.

For a 95% confidence interval,

P(interval contains μ) = 0.95

P(interval does not contain μ) = 0.05

Note: Usually a symmetrical confidence interval is constructed. It is possible to construct one-sided confidence intervals, but these are rarely used.

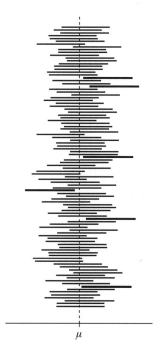

μ

Case 1

We first examine the case of a confidence interval for μ when

- the population variable X is **normally distributed**
- the population variance σ^2 is **known**
- the sample can be **any size**, large or small.

Consider first how to calculate the **confidence limits** (the end-values of the interval) for a **95% confidence interval**. This method can then be adapted for other levels of confidence.

Note that the explanation below shows how the confidence limits are derived. It is useful to know the underlying theory, but in practice you only need to be able to calculate them using the formula.

The theory is based on the distribution of the sample mean, $\overline{X}$.

You know that, for random samples of size n, if $X \sim \text{N}(\mu, \sigma^2)$, then $\overline{X} \sim \text{N}\left(\mu, \dfrac{\sigma^2}{n}\right)$.

See page 99.

Now standardize $\overline{X}$ to get

$$Z = \frac{\overline{X} - \mu}{\dfrac{\sigma}{\sqrt{n}}} \text{ where } Z \sim \text{N}(0, 1)$$

95% confidence interval

To construct a 95% confidence interval, you need the z-values between which the central 95% of the distribution lies. This means that 5% of the distribution lies in the tail ends, with 2.5% (= 0.025) in each tail.

You can see from the diagram that

$$P(Z > z) = 0.025$$

so $\quad P(Z < z) = 0.975$

$$z = \Phi^{-1}(0.975) = 1.96$$

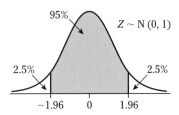

The z-values are ± 1.96,

so $\quad P(-1.96 < Z < 1.96) = 0.95$

i.e. $\quad P\left(-1.96 < \dfrac{\overline{X} - \mu}{\dfrac{\sigma}{\sqrt{n}}} < 1.96\right) = 0.95$

Multiply through by $\dfrac{\sigma}{\sqrt{n}}$:

$$P\left(-1.96 \frac{\sigma}{\sqrt{n}} < \overline{X} - \mu < 1.96 \frac{\sigma}{\sqrt{n}}\right) = 0.95$$

This can be rearranged to give

$$P\left(\overline{X} - 1.96 \frac{\sigma}{\sqrt{n}} < \mu < \overline{X} + 1.96 \frac{\sigma}{\sqrt{n}}\right) = 0.95$$

Now compare this with

$$P(a < \mu < b) = 0.95$$

If the mean of your sample is $\overline{x}$, then

$$a = \overline{x} - 1.96 \frac{\sigma}{\sqrt{n}} \quad \text{and} \quad b = \overline{x} + 1.96 \frac{\sigma}{\sqrt{n}}$$

These are the 95% **confidence limits**.

The **95% confidence limits** are written $\overline{x} \pm 1.96 \dfrac{\sigma}{\sqrt{n}}$.

The **95% confidence interval** is written in the form (a, b),

i.e. $\left(\overline{x} - 1.96 \dfrac{\sigma}{\sqrt{n}}, \ \overline{x} + 1.96 \dfrac{\sigma}{\sqrt{n}}\right)$

If $\bar{x}$ is the mean of a random sample of size n from a normal population with known variance σ^2, then a **95% confidence interval for the population mean μ** is

$$\left(\bar{x} - 1.96\,\frac{\sigma}{\sqrt{n}},\, \bar{x} + 1.96\,\frac{\sigma}{\sqrt{n}}\right)$$

The **midpoint** of the confidence interval is $\bar{x}$.

The **width** of the confidence interval is $2 \times 1.96\,\dfrac{\sigma}{\sqrt{n}}$

The confidence interval is symmetrical about $\bar{x}$.

Example 4.11

The mass of vitamin E in a capsule manufactured by a certain company is normally distributed with standard deviation 0.042 mg. The mean mass of vitamin E in a random sample of 5 capsules was 5.12 mg.

(i)   Calculate a 95% confidence interval for the population mean mass of vitamin E per capsule.

(ii) Calculate the width of the confidence interval.

(iii) 60 random samples of 5 capsules are taken and a 95% confidence interval is calculated for each sample. How many of these intervals would you expect **not** to include μ?

Let X be the mass, in milligrams, of vitamin E in a capsule, where

$$X \sim N(\mu, 0.042^2) \qquad \mu \text{ is unknown, } \sigma = 0.042$$

$\bar{X}$ is the mean mass, in milligrams, of vitamin E in a random sample of 5 capsules, where

$$\bar{X} \sim N\!\left(\mu, \frac{\sigma^2}{n}\right) \qquad \sigma = 0.042,\ n = 5$$

(i)   You are given that $\bar{x} = 5.12$.

95% confidence limits

$$= \bar{x} \pm 1.96\,\frac{\sigma}{\sqrt{n}}$$

$$= 5.12 \pm 1.96\,\frac{0.042}{\sqrt{5}}$$

$$= 5.12 \pm 0.0368\ldots \qquad \longleftarrow \text{ Store 0.0368\ldots in your calculator.}$$

Lower confidence limit $= 5.12 - 0.0368\ldots = 5.0831\ldots$

Upper confidence limit $= 5.12 + 0.0368\ldots = 5.1568\ldots$

So, the 95% confidence interval for μ

$$= (5.08, 5.16) \qquad (3 \text{ s.f.})$$

(ii) width of C. I. $= 2 \times 1.96\,\dfrac{\sigma}{\sqrt{n}}$

$$= 2 \times 0.0368\ldots \qquad \text{from part (i)}$$

$$= 0.07362\ldots$$

$$= 0.0736 \ (3 \text{ s.f.})$$

(iii) On average, 5% of intervals do not include μ, so you would expect 3 intervals not to include μ.

Note: The probability is 0.95 or 95% that the interval (5.08, 5.16) includes, or has trapped, μ. If you took another random sample of the same size, you would probably get a different interval. If you took lots of samples in a similar way, on average 95% of these intervals would include the true population mean μ.

z-values in confidence intervals

In a 95% confidence interval, the z-value is 1.96

In general, the z-value (sometimes referred to as the critical z-value) depends on the confidence level.

90% confidence interval

$$P(Z > z) = 0.05$$

so $P(Z < z) = 0.95$

$$z = \Phi^{-1}(0.95)$$

$$= 1.645$$

You can find z by reading the normal distribution function table (page 199) in reverse. However it is worth checking whether it is one of the values summarised in the **table of critical values** (page 199) as it may be quicker to use this table.

90% of the distribution is in the centre, leaving 10% to be shared between the tails, i.e 5% (= 0.05) in each tail.

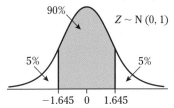

99% confidence interval

$$P(Z > z) = 0.005$$

so $P(Z < z) = 0.995$

$$z = \Phi^{-1}(0.995) = 2.576$$

Note: $\Phi^{-1}(0.995)$ is given in the table of critical values.

1% is shared between the tails, i.e 0.5% (= 0.005) in each tail.

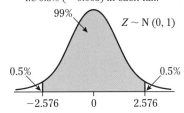

97% confidence interval

$$P(Z > z) = 0.015$$

so $P(Z < z) = 0.985$

$$z = \Phi^{-1}(0.985) = 2.17$$

Note: $\Phi^{-1}(0.985)$ is **not** given in the table of critical values, so use the normal distribution function table in reverse.

3% is shared between the tails, i.e 1.5% (= 0.015) in each tail.

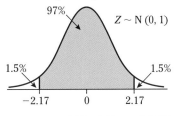

Below is a summary for the 3 most commonly used confidence levels, 90%, 95% and 99%

Level	Confidence limits	Confidence interval	Width
90%	$\bar{x} \pm 1.645 \frac{\sigma}{\sqrt{n}}$	$\left(\bar{x} - 1.645 \frac{\sigma}{\sqrt{n}}, \bar{x} + 1.645 \frac{\sigma}{\sqrt{n}}\right)$	$2 \times 1.645 \frac{\sigma}{\sqrt{n}}$
95%	$\bar{x} \pm 1.96 \frac{\sigma}{\sqrt{n}}$	$\left(\bar{x} - 1.96 \frac{\sigma}{\sqrt{n}}, \bar{x} + 1.96 \frac{\sigma}{\sqrt{n}}\right)$	$2 \times 1.96 \frac{\sigma}{\sqrt{n}}$
99%	$\bar{x} \pm 2.576 \frac{\sigma}{\sqrt{n}}$	$\left(\bar{x} - 2.576 \frac{\sigma}{\sqrt{n}}, \bar{x} + 2.576 \frac{\sigma}{\sqrt{n}}\right)$	$2 \times 2.576 \frac{\sigma}{\sqrt{n}}$

In general, as the *level* of confidence increases, so does the *width* of the confidence interval.

Confidence intervals can be used to comment on claims made about the population mean, as in the following example.

Example 4.12

The masses of sweets produced by a machine are normally distributed with mean μ grams and standard deviation 1.0 grams. A random sample of 65 sweets produced by the machine has a mean mass of 29.6 grams.

(i) Find a 99% confidence interval for μ.

The manufacturer claims that the machine produces sweets with a mean mass of 30 grams.

(ii) Use the confidence interval found in part (i) to draw a conclusion about this claim.

(iii) Another random sample of 65 sweets produced by the machine is taken. This sample gives a 99% confidence interval that leads to a different conclusion from that found in part (ii). Assuming that the value of μ has not changed, explain how this can be possible.

Cambridge Paper 7 Q3 N10(73)

(i) X is the mass, in grams, of a sweet produced by the machine, where $X \sim \mathrm{N}(\mu, \sigma^2)$, with μ unknown and $\sigma = 1.0$

$\overline{X}$ is the mean mass, in grams, of a random sample of 65 sweets, where

$$\overline{X} \sim \mathrm{N}\left(\mu, \frac{\sigma^2}{n}\right) \qquad \sigma = 1.0, n = 65$$

You are given that $\overline{x} = 29.6$

For a 99% C. I., $z = 2.576$

99% confidence limits

$$= x \pm 2.576 \frac{\sigma}{\sqrt{n}}$$

$$= 29.6 \pm 2.576 \frac{1.0}{\sqrt{65}}$$

$$= 29.6 \pm 0.3195\ldots$$

99% confidence interval

$$= (29.28\ldots, 29.91\ldots)$$

$$= (29.3, 29.9) \qquad (3 \text{ s.f.})$$

(ii) The confidence interval does not contain 30, so the manufacturer's claim is not supported.

(iii) As the claim is supported, the confidence interval constructed from the second sample must have contained 30. Confidence intervals will vary, according to the value of the sample mean.

Note: Look again at the computer simulation on page 112.

Note: For a 99% confidence interval, if you were to construct 100 intervals in the same way, on **average**, 99 of them would contain μ and 1 would not. You cannot say with absolute certainty that your particular confidence interval will contain μ.

Case 2

We now examine the case of a confidence interval for μ when

- the distribution of X is **not normal**
- the population variance σ^2 is **known**
- the sample size is **large**.

When the population is not normally distributed, provided the sample size is large ($n \geq 30$, say), you can use the **Central Limit theorem**, where

$$\overline{X} \sim N\left(\mu, \frac{\sigma^2}{n}\right) \text{ approximately}$$ See page 105.

If $\overline{x}$ is the mean of a random sample of size n, where n is **large** ($n \geq 30$), from a non-normal population with known variance σ^2, then an approximate **95% confidence interval for the population mean μ** is

$$\left(\overline{x} - 1.96\,\frac{\sigma}{\sqrt{n}},\ \overline{x} + 1.96\,\frac{\sigma}{\sqrt{n}}\right)$$

Example 4.13

The heights, in cm, of men in a particular district are distributed with mean μ and standard deviation σ. A random sample of 100 men from the district was taken and the mean height, $\overline{x}$, was calculated. A 95% confidence interval for μ was constructed and found to be (177.22, 179.18).

(i)   Calculate the values of $\overline{x}$ and σ.

(ii) Based on this sample, calculate an 80% confidence interval for μ.

(iii) What is the probability that the 80% confidence interval does **not** contain μ?

(iv) Find the probability that the 80% confidence interval contains μ but the 95% confidence interval does not contain μ.

Let X be the height, in centimetres, of a man in this district, where mean μ and variance σ^2 are unknown.

The distribution of X is unknown.

$\overline{X}$ is the mean height, in centimetres, of a random sample of 100 men in the district.

Since the sample size is large, by the Central Limit theorem, $\overline{X} \sim N\left(\mu, \frac{\sigma^2}{n}\right)$ with $n = 100$

(i)   The 95% confidence interval for μ is

$$\left(\overline{x} - 1.96\,\frac{\sigma}{\sqrt{n}},\ \overline{x} + 1.96\,\frac{\sigma}{\sqrt{n}}\right) \text{ with } n = 100 \text{ so } \sqrt{n} = 10$$

Comparing this with (177.22, 179.18),

$$\overline{x} - 1.96\,\frac{\sigma}{10} = 177.22 \qquad\qquad (1)$$

$$\overline{x} + 1.96\,\frac{\sigma}{10} = 179.18 \qquad\qquad (2)$$

(1) + (2)

$$2\overline{x} = 356.4$$

$$\overline{x} = 178.2$$

Alternatively, $\overline{x}$ is the midpoint of the confidence interval, so
$$\overline{x} = \tfrac{1}{2}(177.22 + 179.18) = 178.2$$

(2) − (1) $2 \times 1.96 \times \dfrac{\sigma}{10} = 1.96$

$$\sigma = 5$$

(ii) You know that $\overline{x} = 178.2$, $\sigma = 5$ and $n = 100$.

To find the z-value for an 80% confidence interval, find z such that

For 10% in each tail, look up $\Phi^{-1}(0.9)$.

$$P(Z < z) = 0.9$$

So $z = 1.282$

80% confidence limits

$$= \bar{x} \pm 1.282 \frac{\sigma}{\sqrt{n}}$$

$$= 178.2 \pm 1.282 \frac{5}{\sqrt{100}}$$

$$= 178.2 \pm 0.641$$

80% confidence interval for μ

$$= (177.559, 178.841)$$

The 80% confidence interval is narrower than the 95% confidence interval.

```
                    177.559      178.2       178.841
80% C.I.               ├───────────┼───────────┤

95% C.I.     ├──────────────┼──────────────┤
          177.22          178.2          179.18
```

(iii) $P(80\% \text{ C. I. does not contain } \mu) = 1 - 0.8 = 0.2$

(iv) The confidence intervals are based on the same value of $\bar{x}$ and the 80% C. I. is narrower than the 95% C. I. So, if the 80% C. I. contains μ, then the 95% C. I. will also contain μ.

It is therefore **impossible** for the 80% C. I. to contain μ but the 95% C. I. not to contain μ, so the probability is 0.

Case 3

Finally, we examine the case of a confidence interval for μ when

- the population variance σ^2 is **unknown**
- the sample size is **large**.

When constructing confidence intervals it is often the case that the population variance σ^2 is unknown. Provided that the sample size is large ($n \geqslant 30$, say) it is permissible to use s^2 instead, where s^2 is the **unbiased estimate** of σ^2.

See page 109.

When σ^2 is **unknown**, provided that **n is large** ($n \geqslant 30$), a 95% confidence interval for μ is

$$\left(\bar{x} - 1.96 \frac{s}{\sqrt{n}}, \bar{x} + 1.96 \frac{s}{\sqrt{n}} \right) \qquad \text{where } s^2 = \frac{1}{n-1} \left(\Sigma x^2 - \frac{(\Sigma x)^2}{n} \right)$$

For other confidence levels, calculate the critical value of z, as previously.

Example 4.14

(i) Give a reason why, in carrying out a statistical investigation, a sample rather than a complete population may be used.

(ii) Rose wishes to investigate whether men in her town have a different life-span from the national average of 71.2 years. She looks at government records for her town and takes a random sample of the ages of 110 men who have died recently. Their mean age in years was 69.3 and the unbiased estimate of the population variance was 65.61.

(a) Calculate a 90% confidence interval for the population mean and explain what you understand by this confidence interval.

(b) State with a reason what conclusion about the life-span of men in her town Rose could draw from this confidence interval.

Cambridge Paper 7 Q4 N05

(i) It is less time-consuming to use a sample.

(ii) Let X be the lifespan in years of men in Rose's town, with mean μ and variance σ^2.

Since σ^2 is unknown, use $s^2 = 65.61$

$$\bar{x} = 69.3, s = \sqrt{65.61} = 8.1, n = 110$$

For a 90% confidence interval, $z = 1.645$

90% confidence limits

$$= \bar{x} \pm 1.645 \frac{s}{\sqrt{n}}$$

$$= 69.3 \pm 1.645 \frac{8.1}{\sqrt{110}}$$

$$= 69.3 \pm 1.270...$$

90% confidence interval for μ

$$= (68.029..., 70.570...)$$

$$= (68.0, 70.6) \qquad (3 \text{ s.f.})$$

On average, 90% of intervals constructed in this way will include the population mean μ.

Since the national average lifespan of 71.2 years does not lie in the confidence interval there is evidence that the lifespan of men in Rose's town is significantly different from the national average.

Example 4.15

The fuel consumption of a new model of car is being tested. In one trial, 50 cars chosen at random were driven under identical conditions and the distances x, in kilometres, covered on a litre of fuel were recorded. The results gave the following totals:

$$\Sigma x = 525, \qquad \Sigma x^2 = 5625$$

 (i) Calculate unbiased estimates of the population mean and standard deviation of the distance travelled on a litre of fuel.

 (ii) Calculate a 94% confidence interval for the mean distance travelled on a litre of fuel.

 (iii) Calculate the width of the 94% confidence interval.

A second random sample of 50 cars is taken and a 94% confidence interval is calculated using the sample value obtained.

 (iv) Find the probability that neither confidence interval contains μ.

X is the distance, in kilometres, covered on a litre of fuel, where mean μ and variance σ^2 are unknown.

 (i) Unbiased estimate of μ:

$$\hat{\mu} = \bar{x} = \frac{\Sigma x}{n} = \frac{525}{50} = 10.5$$

Unbiased estimate of σ^2:

$$s^2 = \frac{1}{n-1}\left(\Sigma x^2 - \frac{(\Sigma x)^2}{n}\right)$$

$$= \frac{1}{49}\left(5625 - \frac{525^2}{50}\right)$$

$$= 2.2959$$

(ii) To find the z-value for a 94% confidence interval, find z such that

$$P(Z < z) = 0.97$$

So $z = 1.881$

$\bar{x} = 10.5$, s $= \sqrt{2.2959}$, $n = 50$

94% confidence limits

$$= \bar{x} \pm 1.881\frac{s}{\sqrt{n}}$$

$$= 10.5 \pm 1.881\frac{\sqrt{2.2959}}{\sqrt{50}}$$

$$= 10.5 \pm 0.4030\ldots \quad\longleftarrow\quad \text{Store 0.4030... in your calculator}$$

94% confidence interval for μ

$$= (10.096\ldots, 10.903\ldots)$$

$$= (10.1, 10.9) \quad (3\text{ s.f.})$$

(iii) width $= 2 \times 1.881\frac{s}{\sqrt{n}}$

$\quad\quad = 2 \times 0.4030\ldots$ $\quad$ from part (ii)

$\quad\quad = 0.8061\ldots$

$\quad\quad = 0.806\,\text{km}$ (3 s.f.)

You could find the difference between the confidence limits found in part (ii), but take care not to use the rounded figures as these will lead to inaccuracy.

(iv) P(first C. I. contains μ) $= 0.94$

P(first C. I. does not contain μ) $= 1 - 0.94 = 0.06$

Similarly,

$\quad\quad$ P(second C. I. does not contain μ) $= 0.06$

So, P(neither contains μ) $= 0.06 \times 0.06 = 0.0036$

Determination of sample size

Example 4.16

The result, X, of a stress test on a metal is known to be normally distributed with mean μ and standard deviation 1.3

It is required to have a 95% confidence interval for μ with a width of less than 2.0

Find the smallest number of tests needed to achieve this.

$X \sim N(\mu, 1.3^2)$ $\quad\quad$ μ is unknown, $\sigma = 1.3$

For samples of size n, $\bar{X} \sim N\left(\mu, \frac{1.3^2}{n}\right)$

Width of 95% confidence interval

$$= 2 \times 1.96\frac{\sigma}{\sqrt{n}}$$

$$= 2 \times 1.96\frac{1.3}{\sqrt{n}} \quad\quad (1)$$

$$= \frac{5.096}{\sqrt{n}}$$

The width of the interval must be less than 2,

so $\dfrac{5.096}{\sqrt{n}} < 2$

$\qquad 5.096 < 2\sqrt{n}$ (2)

$\qquad \sqrt{n} > \dfrac{5.096}{2}$

$\qquad \sqrt{n} > 2.584$

Square both sides:

$\qquad n > 6.67\ldots$

So the smallest number of tests needed is 7.

Now suppose that in Example 4.16 the 95% confidence interval for μ must have a width less than 1, instead of less than 2.

To achieve a width less than 1, will the sample size be larger or smaller?

Consider equation (2) in Solution 4.16.

For a width less than 1, this becomes

$\qquad 5.096 < 1 \times \sqrt{n}$

$\qquad \sqrt{n} > 5.096$

$\qquad n > 25.96\ldots$

So the smallest sample size needed for a width less than 1 is 26. This is larger than the sample size of 7 needed for a width less than 2.

In general, for a **given confidence level**,

– the smaller the interval width, the larger the sample size required.

Now suppose that, in Example 4.16, the width of the interval must still be less than 2, but the confidence level is increased to 99%

To achieve this, will the sample size be larger or smaller?

The calculation needed to find the width is similar to (1) in Solution 4.16, but replace 1.96 with the critical value of z for a 99% confidence interval.

For a 99% confidence interval, $z = 2.576$, so width $= 2 \times 2.576 \dfrac{1.3}{\sqrt{n}} = \dfrac{6.6976}{\sqrt{n}}$

You want $\dfrac{6.6976}{\sqrt{n}} < 2$

$\qquad 6.6976 < 2\sqrt{n}$

$\qquad \sqrt{n} > 3.3488$

$\qquad n > 11.21\ldots$

For a 99% confidence interval, the smallest sample size needed is 12, whereas for a 95% confidence interval the smallest sample size is 7.

In general, for a **given interval width**,

– the greater the confidence level, the larger the sample size required.

Exercise 4e

1 A student, studying the height of a particular plant, knows that it follows a normal distribution with mean μ cm and standard deviation 1.4 cm. He selects 15 plants at random, measures their heights and calculates that the mean height of the sample is 12.2 cm.

   (i) Calculate a 90% confidence interval for μ.

   (ii) Calculate the width of this interval.

2 A certain type of tennis ball is known to have a height of bounce which is normally distributed with standard deviation 2 cm. A sample of 60 tennis balls is tested and the mean height of bounce of the sample is 140 cm.

   (i) Find a 95% confidence interval for the mean height of bounce of this type of tennis ball.

   (ii) State any assumptions made in calculating your interval.

   (iii) The manufacturer claims that the mean height of bounce is 141 cm. Use the confidence interval found in part (i) to draw a conclusion about this claim.

3 A random sample of 6 items taken from a normal population with mean μ centimetres and variance 4.5 cm² gave the following values, in centimetres:

   12.9 13.2 14.6 12.6 11.3 10.1

   (i) Calculate the mean of the sample.

   (ii) Find the 95% confidence interval for μ.

   (iii) Find the width of the 95% confidence interval.

4 A factory produces cans of fruit whose masses are normally distributed with standard deviation 18 g. A random sample of 25 cans of fruit is found to have a mean mass of 458 g.

   (i) Calculate the 99% confidence interval for the population mean mass of cans of fruit produced at the factory.

   (ii) Explain what the interval means.

   (iii) Would the interval be wider if a 90% confidence interval was calculated? Explain your reasoning.

5 A random sample of 100 observations from a normal population with mean μ gave the following data:

   $\Sigma x = 8200, \qquad \Sigma x^2 = 686\,800$

   (i) Find a 98% confidence interval for μ.

   (ii) Find a 99% confidence interval for μ.

   (iii) Would your answers have been different if the population was not normal? Explain your answer.

6 The height x, in centimetres, of each man in a random sample of 200 men living in a certain city was measured. The following results were obtained.

   $\Sigma x = 35\,050, \qquad \Sigma x^2 = 6\,163\,109$

   (i) Calculate unbiased estimates of the mean and variance of the heights of men living in the city.

   (ii) Calculate an approximate 90% confidence interval for the mean height of men living in the city.

   (iii) Explain the use of the Central Limit theorem in your calculations for part (ii).

7 The time T, in minutes, to complete a particular test was measured for a random sample of 65 students.

   The sample mean was 53.2 min and an unbiased estimate of the population variance was 37.4 min².

   (i) Determine a 99% confidence interval for the mean time taken by students to complete the test.

   (ii) A lecturer claims that the mean time taken by students to complete the test is 50 min. Use the confidence interval found in part (i) to draw a conclusion about this claim.

8 A firm has 763 employees. A researcher is investigating pulse rates and wishes to select a random sample of 80 employees to take part in his research.

   (i) Explain briefly how a suitable random sample could be selected.

(ii) A suitable random sample of 80 employees is selected. The researcher asks these employees to measure their pulse rates when they wake up in the morning. The mean rate for the employees in the sample is 69 beats. An unbiased estimate of the standard deviation of all the employees in the firm is 4 beats.

(a) Calculate a 97% confidence interval for the mean pulse rate of all the employees at the firm.

(b) Find the width of this interval.

The researcher decides to take a second random sample of 100 employees. Based on their mean rate she constructs a 95% confidence interval for the mean pulse rate of all the employees in the firm.

(c) Find the probability that just one of the two confidence intervals constructed by the researcher contains the true population mean.

9 A factory produces bags of flour. A random sample of 150 bags of flour is taken from the production line. The mean mass of the bags in the sample is 748 g and the standard deviation is 3.6 g.

Calculate a 98% confidence interval for the mean mass of bags of flour produced in the factory.

10 The mean lifetime of a particular type of light bulb is μ hours.

(i) A 95% confidence interval for μ was calculated and the confidence limits were 1023.3 h and 1101.7 h. The interval was based on the results of a random sample of 36 light bulbs. Calculate a 99% confidence interval for μ.

(ii) 40 random samples of 36 light bulbs are taken and a 90% confidence interval for μ is calculated for each sample. Find the expected number of intervals that contain μ.

11 A random sample of 60 loaves is taken from a population whose masses are normally distributed with mean μ grams and standard deviation 10 grams.

(i) Find the width of a 95% confidence interval for μ based on this sample.

(ii) Find the confidence level of a 95% confidence interval having the same width as in part (i), but based on a random sample of 40 loaves.

12 The height, in centimetres, of a particular type of shrub is known to be normally distributed with mean μ and standard deviation 12.5

A random sample of n trees is selected and their heights measured.

It is required to have a 90% confidence interval for μ with a total width of less than 6.0 cm. Find the least value of n needed to achieve this.

CONFIDENCE INTERVALS FOR p, THE POPULATION PROPORTION

Suppose you want to know p, the proportion of 'successes' in a particular population. An **unbiased estimate of p** is given by p_s, the proportion of successes in a random sample. If there are x successes in a random sample of size n, then $p_s = \dfrac{x}{n}$

The sample proportion p_s is also used to obtain a **confidence interval for p**. This is obtained from the distribution of the sample proportion, P_s.

Distribution of the sample proportion P_s

Suppose a random sample of n observations is taken from a population in which the proportion of successes is p and the proportion of failures is $q = 1 - p$.

If X is the **number of successes** in the sample, then X follows a **binomial distribution**, i.e. $X \sim \text{B}(n, p)$ with mean np and variance npq.

If X is the **number of successes** in the sample, then X follows a **binomial distribution**, i.e. $X \sim B(n, p)$ with mean $E(X) = np$ and variance $\text{Var}(X) = npq$.

Now consider P_s, the **proportion of successes** in the sample, where $P_s = \dfrac{X}{n}$.

Using the results from chapter 2, it is possible to work out the mean and variance of P_s as follows.

$$E(P_s) = E\left(\frac{X}{n}\right)$$
$$= E\left(\frac{1}{n}X\right)$$
$$= \frac{1}{n}E(X) \qquad E(aX) = aE(X) \text{ (page 26)}$$
$$= \frac{1}{n} \times np$$
$$= p$$

$$\text{Var}(P_s) = \text{Var}\left(\frac{X}{n}\right)$$
$$= \frac{1}{n^2}\text{Var}(X) \qquad \text{Var}(aX) = a^2\text{Var}(X) \text{ (page 26)}$$
$$= \frac{1}{n^2} \times npq$$
$$= \frac{pq}{n}$$

You will not need these proofs in the exam

So, the distribution of P_s has mean p and variance $\dfrac{pq}{n}$

When **n is large**, the normal approximation to the binomial distribution can be used and the distribution of P_s is **approximately normal**.

Provided the sample size n is **large** ($n \geqslant 30$), the distribution of P_s is **approximately normal**, where

$$P_s \sim N\left(p, \frac{pq}{n}\right) \qquad \text{with } q = 1 - p$$

The larger the sample size n, the better the approximation.

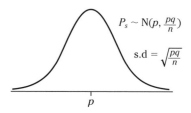

$P_s \sim N(p, \frac{pq}{n})$

$\text{s.d} = \sqrt{\frac{pq}{n}}$

The standard deviation of P_s is known as the **standard error of proportion**.

The standard deviation $\sqrt{\dfrac{pq}{n}}$ is needed to calculate the confidence limits of the confidence interval.

The difficulty, however, is that you do not know p.

To overcome this, p_s is used, giving an approximate value of

the standard deviation of $\sqrt{\dfrac{p_s q_s}{n}}$, where $q_s = 1 - p_s$

If you know p, you do not need to set up a confidence interval for it!

Confidence intervals for p

As with the confidence interval for the population mean, an interval (a, b) is constructed that has a specified probability of including the true population proportion p.

The confidence limits are $p_s \pm z\sqrt{\dfrac{p_s q_s}{n}}$, where the value of z depends on the level of confidence.

For example, for a 95% confidence interval, $z = 1.96$

95% confidence interval

Provided n is **large** ($n \geqslant 30$), an **approximate** 95% confidence interval for p is

$$\left(p_s - 1.96\sqrt{\frac{p_s q_s}{n}}, \; p_s + 1.96\sqrt{\frac{p_s q_s}{n}}\right)$$

where p_s is the proportion of successes in the sample.

The **midpoint** of the confidence interval is p_s.

The **width** of the 95% confidence interval

is $2 \times 1.96\sqrt{\dfrac{p_s q_s}{n}}$.

$p_s - 1.96\sqrt{\frac{p_s q_s}{n}} \qquad p_s \qquad p_s + 1.96\sqrt{\frac{p_s q_s}{n}}$

width

$2 \times 1.96\sqrt{\frac{p_s q_s}{n}}$

Below is a summary for the three most commonly used confidence intervals, 90%, 95% and 99%

Level	Confidence limits	Confidence interval	Width
90%	$p_s \pm 1.645 \sqrt{\dfrac{p_s q_s}{n}}$	$\left(p_s - 1.645 \sqrt{\dfrac{p_s q_s}{n}}, p_s + 1.645 \sqrt{\dfrac{p_s q_s}{n}}\right)$	$2 \times 1.645 \sqrt{\dfrac{p_s q_s}{n}}$
95%	$p_s \pm 1.96 \sqrt{\dfrac{p_s q_s}{n}}$	$\left(p_s - 1.96 \sqrt{\dfrac{p_s q_s}{n}}, p_s + 1.96 \sqrt{\dfrac{p_s q_s}{n}}\right)$	$2 \times 1.96 \sqrt{\dfrac{p_s q_s}{n}}$
99%	$p_s \pm 2.576 \sqrt{\dfrac{p_s q_s}{n}}$	$\left(p_s - 2.576 \sqrt{\dfrac{p_s q_s}{n}}, p_s + 2.576 \sqrt{\dfrac{p_s q_s}{n}}\right)$	$2 \times 2.576 \sqrt{\dfrac{p_s q_s}{n}}$

Remember that n must be large.

Since a continuous distribution (normal) has been used as an approximation to a discrete distribution (binomial), a continuity correction should be used. However, **continuity corrections are usually omitted when calculating confidence intervals**.

The confidence interval for p is approximate because:

- the distribution of P_s is approximately normal
- the population standard deviation $\sqrt{\dfrac{pq}{n}}$ has been estimated using $\sqrt{\dfrac{p_s q_s}{n}}$

Example 4.17

A survey was conducted to find the proportion of people owning DVD players. It was found that 203 out of a random sample of 278 people owned a DVD player.

(i) Calculate a 97% confidence interval for the true proportion of people who own a DVD player.

A second survey to find the proportion of people owning DVD players was conducted at 10 o'clock on a Thursday morning in a shopping centre.

(ii) Give one reason why this is not a satisfactory sample.

Cambridge Paper 7 Q3 N06

(i) Let p be the population proportion of people who own a DVD player.

$$p_s = \frac{203}{278} = 0.7302 \text{ (4 s.f.)}, \quad q_s = 1 - \frac{203}{278} = 0.2698 \text{ (4 s.f.)},$$

$$n = 278$$

> For greater accuracy, you could leave p_s and q_s as fractions, but it is often easier to work in decimals. As the final answer is given to 3 significant figures, you must work to at least 4 significant figures throughout.

For 97% confidence interval,

$$P(Z < z) = 0.985$$

$$z = \Phi^{-1}(0.985) = 2.17$$

97% confidence limits

$$= p_s \pm 2.17 \sqrt{\frac{p_s q_s}{n}}$$

$$= 0.7302 \pm 2.17 \sqrt{\frac{0.7302 \times 0.2698}{278}}$$

$$= 0.7302 \pm 0.0577\ldots \quad \longleftarrow \text{ Store } 0.0577\ldots \text{ in memory}$$

97% confidence interval for p

$$= (0.672, 0.788) \quad \text{(3 s.f.)}$$

(ii) This sample would not be representative of the whole population as people selected in the shopping centre at this time of day are more likely to include people who are unemployed, retired or looking after young children. Those who are at work at this time have not had the opportunity to be selected.

Example 4.18

A survey of a random sample of n people found that 61 of them read *The Reporter* newspaper. A symmetric confidence interval for the true population proportion, p, who read *The Reporter* is $0.1993 < p < 0.2887$

(i) Find the mid-point of this confidence interval and use this to find the value of n.

(ii) Find the confidence level of this confidence interval.

Cambridge Paper 7 Q3 J05

(i) For the interval $0.1993 < p < 0.2887$,

midpoint $= \frac{1}{2}(0.1993 + 0.2887) = 0.244$

The confidence interval for p is $\left(p_s - z\sqrt{\dfrac{p_s q_s}{n}}, p_s + z\sqrt{\dfrac{p_s q_s}{n}}\right)$, where z is the critical value for the level of confidence.

The midpoint is p_s, the sample proportion, so $p_s = 0.244$

$$
\begin{array}{ccc}
0.1993 & & 0.2887 \\
\hline
p_s - z\sqrt{\frac{p_s q_s}{n}} & p_s & p_s + z\sqrt{\frac{p_s q_s}{n}}
\end{array}
$$

But you are given that $p_s = \dfrac{61}{n}$,

so $\quad 0.244 = \dfrac{61}{n}$

$\qquad n = \dfrac{61}{0.244}$

$\qquad\quad = 250$

(ii) *To find the confidence level, first find the value of z.*

Width of C. I. $= 2z\sqrt{\dfrac{p_s q_s}{n}}$ $\qquad\qquad q_s = 1 - 0.244 = 0.756$

$\qquad\qquad\quad = 2 \times z \times \sqrt{\dfrac{0.244 \times 0.756}{250}}$

$\qquad\qquad\quad = 0.0543\ldots \times z$ $\qquad$ Store 0.0543... in your calculator memory.

But, using the given confidence limits,

width of C. I. $= 0.2887 - 0.1993 = 0.0894$

So, $\quad 0.0543\ldots \times z = 0.0894$

$\qquad\qquad z = \dfrac{0.0894}{0.0543\ldots} = 1.646$

$\quad P(|Z| < 1.646) = 2\Phi(1.646) - 1$

$\qquad\qquad\qquad = 2 \times 0.95 - 1$

$\qquad\qquad\qquad = 0.90$

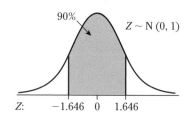

The confidence level is 90%.

Example 4.19

(i) Explain what is meant by the term 'random sample'.

In a random sample of 350 food shops it was found that 130 of them had Special Offers.

(ii) Calculate an approximate 95% confidence interval for the proportion of all food shops with Special Offers.

(iii) Estimate the size of a random sample required for an approximate 95% confidence interval for this proportion to have a width of 0.04.

Cambridge Paper 7 Q3 N07

(i) A random sample of size n is a sample chosen such that
- every member of the population has an equal chance of being selected
- all possible samples of size n have an equal chance of being selected.

(ii) Let p be the proportion of all food shops with special offers.

$$p_s = \frac{130}{350}, \qquad q_s = 1 - \frac{130}{350} = \frac{220}{350}, \qquad n = 350$$

95% confidence limits

$$= p_s \pm 1.96 \sqrt{\frac{p_s q_s}{n}}$$

$$= \frac{130}{350} \pm 1.96 \sqrt{\frac{\frac{130}{350} \times \frac{220}{350}}{350}}$$

$$= \frac{130}{350} \pm 0.05062\ldots$$

95% confidence interval for p

$$= (0.321, 0.422) \qquad \text{(3 s.f.)}$$

(iii) width $= 2 \times 1.96 \sqrt{\dfrac{p_s q_s}{n}}$

$$= 2 \times 1.96 \sqrt{\frac{\frac{130}{350} \times \frac{220}{350}}{n}}$$

You want the width to be 0.04,

so $\quad 2 \times 1.96 \sqrt{\dfrac{0.23346\ldots}{n}} = 0.04$

$$\sqrt{\frac{0.23346\ldots}{n}} = 0.01020\ldots$$

Square both sides

$$\frac{0.23346\ldots}{n} = (0.01020\ldots)^2$$

$$n = \frac{0.23346\ldots}{(0.01020\ldots)^2}$$
You must give a whole-number answer for n.

$$= 2242.24$$

So a sample size of 2243 is required.

Exercise 4f

1 In a survey of a random sample of 250 schoolchildren in a certain region, 170 children had at least one pet.

(i) Calculate an approximate 95% confidence interval for the proportion of children in the region who have at least one pet.

(ii) Find the width of the confidence interval.

(iii) Explain why the interval is approximate.

2 In order to assess the probability of a successful outcome, an experiment was performed 200 times. The number of successful outcomes was 72.

(i) Calculate an approximate 95% confidence interval for p, the probability of a successful outcome.

(ii) Calculate the width of an approximate 99% confidence interval for p.

3 Tubes of Stay Clean toothpaste are advertised as containing 75 ml of toothpaste. In a random sample of 80 tubes, 7 were found to be underweight.

Calculate an approximate 92% confidence interval for the proportion of all tubes of Stay Clean toothpaste that are underweight.

4 A survey was undertaken of the use of the internet by residents in a large city. In a random sample of 150 residents, 49 logged on to the internet at least once a day.

(i) Calculate an approximate 90% confidence interval for p, the proportion of residents in the city that log on to the internet at least once a day.

(ii) A total of 100 similar surveys are carried out and the 90% confidence interval calculated for each survey. State the expected number of intervals that include p.

5 Recruits are issued with boots when they join the army. The last 50 pairs of boots issued were the following sizes:

```
  8   9   8  10  11   8   7  12  12   9
  9   8  11   8   9   7  11  12  11  10
  9  10  10  10   8   8   7  12   9   9
 10  13   7   8   9   9  10  10   8  12
  9   9  10  10  11  12   9   9  10   9
```

(i) Find the proportion in the sample requiring size 9.

(ii) Assuming that these recruits can be regarded as a random sample of all recruits, calculate an approximate 94% confidence interval for the proportion, p, of all recruits requiring size 9 boots.

(iii) Explain why the interval is approximate.

6 In a market research survey, 25 people out of a random sample of 100 in a certain town said that they regularly used a particular brand of soap.

(i) Calculate an approximate 97% confidence interval for the proportion of people in the town who regularly use this brand of soap.

(ii) Give a reason why this interval may not contain the true proportion of people in the town who regularly used a particular brand of soap.

7 Out of 248 cars parked in a car park, 72 were fitted with an anti-theft device on the steering wheel.

Assuming that the cars form a random sample of all cars, calculate an approximate 95% confidence interval for the population proportion of cars fitted with an anti-theft device on the steering wheel.

8 A survey was carried out to assess the response to a new 'healthy eating' menu introduced into all the schools in a particular region. A random sample of 200 schoolchildren was selected from the region and it was found that 84 children approved of the new menu.

(i) Calculate an approximate 95% confidence interval for the population proportion, p, who approve of the new menu.

(ii) Find the width of the 95% confidence interval.

(iii) Estimate the size of a random sample required for an approximate 95% confidence interval for this proportion to have a width of 0.06

9 A college principal decides to consult the students about a proposed change in the times of lectures. She finds that 57 students in a random sample of 80 students are in favour of the change.

(i) Calculate an approximate 90% confidence interval for the proportion of students who are **not** in favour of the change.

(ii) State the effect on the width of such a confidence interval when the confidence level is increased.

10 (i) In an opinion poll, 2000 people were interviewed and 527 said that they preferred white chocolate to milk chocolate. Stating any necessary assumptions, calculate an approximate 95% confidence interval for the proportion of the population who prefer white chocolate.

(ii) The α% confidence interval for the proportion preferring white chocolate, based on a sample of size 500, is (0.2278, 0.2922).

(a) Find the proportion of people in the sample of 500 who preferred white chocolate.

(b) Find the value of α.

11 In an opinion poll carried out before a local election, 501 people in a random sample of 925 voters declare that they will vote for a particular one of the two candidates contesting the election. Calculate an approximate 99% confidence interval for the proportion of all voters who will vote for this candidate.

12 The proportion of bruised apples in a large consignment is denoted by p. A sample of 100 apples is examined and 9 apples are found to be bruised.

(i) Give an assumption under which it would be valid to calculate an approximate confidence interval for p.

(ii) Given that the assumption in part (i) is justified, calculate an approximate 90% confidence interval for p.

13 A consumer group wishes to estimate the proportion, p, of pre-packed cheese sandwiches whose salt content is greater than that stated on the label. A random sample of 40 pre-packed cheese sandwiches was tested and 9 packages were found to contain more salt than stated on the label.

(i) Estimate the number of packages that would have to be tested in order that a 95% confidence interval for p should have a width of approximately 0.1

(ii) Would the number of packages to be tested be larger or smaller than the answer in part (i) if the confidence level were changed to 90%?

Summary

Random samples

A random sample of size n is a sample chosen such that:
- every member of the population has an equal chance of being selected
- all possible samples of size n have an equal chance of being selected.

Distribution of the sample mean

If a large number of random samples, each of size n, are taken from a population X, with mean μ and variance σ^2, and the sample mean, $\bar{x}$, calculated for each sample, then these form the distribution of the sample mean, $\overline{X}$.

When X is normally distributed:

If $\quad X \sim N(\mu, \sigma^2)$, then $\overline{X} \sim N\left(\mu, \dfrac{\sigma^2}{n}\right)$, for all values of n.

When X is not normally distributed:

By the **Central Limit theorem**,

$$\overline{X} \sim N\left(\mu, \frac{\sigma^2}{n}\right),\text{ approximately, provided } \textbf{\textit{n} is large,} (n \geqslant 30).$$

The larger the sample size n, the better the approximation.

Unbiased estimates of the population mean and variance

$$\hat{\mu} = \overline{x} = \frac{\Sigma x}{n}, \qquad s^2 = \frac{1}{n-1}\left(\Sigma x^2 - \frac{(\Sigma x)^2}{n}\right)$$

The notation $\hat{\sigma}^2$ is sometimes used instead of s^2.

95% confidence interval for μ

The midpoint of the confidence interval is $\overline{x}$

Conditions	95% confidence interval	Width
Normal population – known variance – any size sample	$\left(\overline{x} - 1.96\,\frac{\sigma}{\sqrt{n}}, \overline{x} + 1.96\,\frac{\sigma}{\sqrt{n}}\right)$	$2 \times 1.96\,\frac{\sigma}{\sqrt{n}}$
Non-normal population – known variance – large sample $(n \geqslant 30)$	Approximate confidence interval: $\left(\overline{x} - 1.96\,\frac{\sigma}{\sqrt{n}}, \overline{x} + 1.96\,\frac{\sigma}{\sqrt{n}}\right)$	$2 \times 1.96\,\frac{\sigma}{\sqrt{n}}$
Any population – unknown variance – large sample $(n \geqslant 30)$	Approximate confidence interval: $\left(\overline{x} - 1.96\,\frac{s}{\sqrt{n}}, \overline{x} + 1.96\,\frac{s}{\sqrt{n}}\right)$	$2 \times 1.96\,\frac{s}{\sqrt{n}}$

Distribution of sample proportion P_s

Provided the sample size n is **large** $(n \geqslant 30)$, the distribution of P_s is **approximately normal**, where

$$P_s \sim N\left(p, \frac{pq}{n}\right) \qquad \text{with } q = 1 - p$$

95% confidence interval for p

p_s is the proportion in the sample, $q_s = 1 - p_s$.

The midpoint of the confidence interval is p_s.

Condition	95% confidence interval	Width
Sample size large $(n \geqslant 30)$	Approximate confidence interval: $\left(p_s - 1.96\,\sqrt{\frac{p_s q_s}{n}}, p_s + 1.96\,\sqrt{\frac{p_s q_s}{n}}\right)$	$2 \times 1.96\,\sqrt{\frac{p_s q_s}{n}}$

Mixed Exercise 4

1 Bars of chocolate have weights that are distributed normally with mean 105.0 g and standard deviation 1.3 g. A random sample of 20 packets is taken. Find the probability that the mean weight of the packets in the sample is more than 105.6 g.

2 The random variable X is distributed Po(4.1). Find the probability that the mean of a random sample of 50 observations of X is less than 4.5

3 The scores, X_1 and X_2, in papers 1 and 2 of an examination are normally distributed with means 24.3 and 31.2 respectively and standard deviations 3.5 and 3.1 respectively. The final mark for each candidate is found by calculating $2X_1 + 1.5X_2$. Find the probability that a random sample of 8 candidates will have a mean final mark of less than 60.

4 In athletics matches the triple jump event consists of a hop, followed by a step, followed by a jump. The lengths covered by Albert in each part are independent normal variables with means 3.5 m, 2.9 m, 3.1 m and standard deviations 0.3 m, 0.25 m, 0.35 m respectively. The length of the triple jump is the sum of the three parts.

   (i) Find the mean and standard deviation of the length of Albert's triple jumps.

   (ii) Find the probability that the mean of Albert's next four triple jumps is greater than 9 m.
   Cambridge Paper 7 Q2 J04

5 When Sunil travels from his home in England to visit his relatives in India, his journey is in four stages. The times, in hours, for the stages have independent normal distributions as follows:

Bus from home to the airport: N(3.75, 1.45)

Waiting in the airport: N(3.1, 0.785)

Flight from England to India: N(11, 1.3)

Car in India to relatives: N(3.2, 0.81)

   (i) Find the probability that the flight time is shorter than the total time for the other three stages.

   (ii) Find the probability that, for 6 journeys to India, the mean time waiting in the airport is less than 4 hours.
   Cambridge Paper 7 Q6 J09(71)

6 An engineering plant produces steel sheets whose weights are normally distributed with standard deviation 2.4 kg. A random sample of 36 sheets had a mean weight of 31.4 kg.

   (i) Find a 99% confidence interval for the mean weight of the steel sheets produced at the plant.

   (ii) Find the width of a 90% confidence interval based on the same sample.

7 A consumer group, interested in the mean fat content of a particular type of sausage, takes a random sample of 20 sausages and sends them away to be analysed. The percentage of fat in each sausage is as follows:

26   27   28   28   28   29   29   30   30   31
32   32   32   33   33   34   34   34   35   35

Assume that the percentage of fat is normally distributed with mean μ, and that the standard deviation is known to be 3.

   (i) Calculate a 98% confidence interval for the population mean percentage of fat.

   (ii) The manufacturer claims that the mean percentage of fat in sausages of this type is 30. Use your answer to part (i) to determine whether the consumer group should accept this claim.
   Cambridge Paper 7 Q3 J03

8 Packets of cat food are filled by a machine.

   (i) In a random sample of 10 packets, the weights, in grams, of the packets were as follows:

374.6   377.4   376.1   379.2   371.2
375.0   372.4   378.6   377.1   371.5

   Find unbiased estimates of the population mean and variance.

   (ii) In a random sample of 200 packets, 38 were found to be underweight. Calculate a 96% confidence interval for the population proportion of underweight packets.
   Cambridge Paper 7 Q4 J04

9 The daily takings \$$x$, for a shop were noted on 30 randomly chosen days. The takings are summarised by

$$\Sigma x = 31\,500, \qquad \Sigma x^2 = 33\,141\,816.$$

(i) Calculate unbiased estimates of the population mean and variance of the shop's daily takings.

(ii) Calculate a 98% confidence interval for the mean daily takings.

The mean daily takings for a random sample of n days is found.

(iii) Estimate the value of n for which it is approximately 95% certain that the sample mean does not differ from the population mean by more than \$6.
Cambridge Paper 7 Q6 J07

10 The weights in grams of oranges grown in a certain area are normally distributed with mean μ and standard deviation σ. A random sample of 50 of these oranges was taken, and a 97% confidence interval for μ based on this sample was (222.1, 232.1).

(i) Calculate unbiased estimates of μ and σ^2.

(ii) Estimate the sample size that would be required in order for a 97% confidence interval for μ to have width 8.
Cambridge Paper 7 Q2 J09(71)

11 A random sample of n people were questioned about their internet use. 87 of them had a high-speed internet connection. A confidence interval for the population proportion having a high-speed internet connection is $0.1129 < p < 0.1771$

(i) Write down the mid-point of this confidence interval and hence find the value of n.

(ii) This interval is an α% confidence interval. Find α.
Cambridge Paper 7 Q2 J10(71)

12 108 out of a random sample of 200 people in Nester said they support Nester Rovers.

(i) Calculate a 94% confidence interval for the true proportion of people in Nester who support the team.

(ii) The same sample was used to calculate an α% confidence interval whose width is less than 0.1. Find the largest possible value of α.

13 In a random sample of 400 carpet shops, it was discovered that 136 of them sold carpets at below the list prices recommended by the manufacturer.

(i) Estimate the proportion of all carpet shops selling below list price.

(ii) Calculate an approximate 90% confidence interval for the proportion of shops that sell below list price and explain briefly what this means.

(iii) What size sample would have to be taken in order to estimate the proportion to within ± 0.02, with 90% confidence?

14 Each of a random sample of 50 UK one-pound coins was weighed and their masses, x grams, are summarised by

$$\Sigma x = 474.51, \qquad \Sigma x^2 = 4503.8276$$

(i) Calculate unbiased estimates of the mean and variance of the masses of UK one-pound coins.

(ii) Find an approximate 90% confidence interval for the mean mass (in grams) of all UK one-pound coins.

(iii) Estimate the size of a random sample of UK one-pound coins that would be required to give a 95% confidence interval whose width is half that of the interval calculated in (ii).

(iv) It was found later that the scales were consistently underweighing by 0.05 grams. State which of the answers in part (ii) and part (iii) should be amended and which should not and give the amended values.

5 Hypothesis tests 1: Discrete variables

In this chapter you will learn about

- the language of hypothesis testing
  - null and alternative hypotheses
  - one-tail and two-tail tests
  - significance level
  - critical (rejection) region
  - test statistic
  - rejection rule
- formulating hypotheses and carrying out a hypothesis test
  - for the parameter p of a binomial distribution
  - for the mean λ of a Poisson distribution
- Type I and Type II errors

HYPOTHESIS TEST FOR A BINOMIAL PROPORTION p

Sid says that he has psychic powers and can read people's thoughts.

To test this claim, a volunteer from the audience sits on the stage while Sid sits in a separate room off stage. The volunteer chooses a card from a well-shuffled pack and concentrates on the card for five seconds. At the same time, Sid writes down the suit of the card, either clubs, diamonds, hearts or spades. The card is replaced in the pack, the pack is shuffled and another card drawn. The procedure is repeated until 20 cards have been drawn.

There are four suits, so if Sid guesses the answer he has a one in four chance of writing down the correct suit. If he isn't guessing, you would expect him to get more than one in four correct. So in the 20 trials you would expect him to get more than 5 correct.

If he gets at most 5 correct out of 20, you would definitely say that he is just guessing. If he gets as many as 19 or 20 correct you would have no hesitation in saying that he could read people's thoughts. But what about other values? If he gets 12 correct answers, would this be very unusual? What would you say if he got 10 correct? What about 8 correct?

Somehow you have to decide on a cut-off point. This would be the smallest value of c you could find such that the **probability** of getting c **or more correct answers** would be **very small** and it would be considered a rare event to get c or more correct answers.

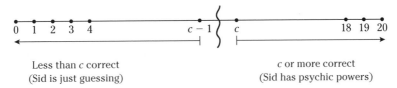

To decide the value of c, you could just choose a number that seemed reasonable. However, if you carry out a **hypothesis (or significance) test** you will be able to back up your conclusion with statistical theory.

Is Sid guessing?

Suppose that X is the number of correct answers Sid writes down for the suits of the 20 cards. If you assume that Sid is guessing, the probability that he writes down the correct suit is 0.25

The experiment is performed 20 times. As the card is replaced and the pack well-shuffled each time, the trials are independent, so there are 20 independent trials, each with a probability of 0.25 of success.

If you assume that Sid is guessing, then X can be modelled by a binomial distribution with $n = 20$ and $p = 0.25$,

i.e. $X \sim B(20, 0.25)$ with $P(X = r) = \binom{n}{r} \times 0.25^r \times 0.75^{n-r}$ S1 Chapter 3, Binomial distribution

Using binomial probabilities, you will find that the probability of Sid getting 20 correct answers if he is guessing is extremely small (0.0000000000009). In fact, the probability of Sid getting 16 or more correct answers is still very small indeed (0.000000386...).

Investigating further, you will find the following probabilities.

Number of correct answers	Probability	
13 or more	$0.00018... \approx 0.02\%$	If Sid is guessing, it would be *almost impossible* for him to give 13 or more correct answers. So if, for example, he gives 13 correct answers, you would certainly conclude that he is able to read people's thoughts.
12 or more	$0.00093... \approx 0.09\%$	Getting 12 or more correct answers is also a *very rare* event.
11 or more	$0.0039... \approx 0.4\%$	Getting 11 or more correct answers is a *rare* event.
10 or more	$0.013... \approx 1\%$	It would be *very unlikely* for Sid to give 10 or more correct answers, but you would expect it to happen on about one in every 100 occasions.
9 or more	$0.040... \approx 4\%$	It would still be *unlikely* for Sid to give 9 or more correct answers.
8 or more	$0.101... \approx 10\%$	This probability is not that small. If Sid is guessing, on 10% of occasions he could give 8 or more correct answers.

You have to make a decision about the value of the probability that is considered to imply an unlikely or rare event. This probability is called the **significance level** of the test. As a guide, events that have a probability of less than 5% are generally regarded as *unlikely* and events having a probability of less than 1% are generally regarded as *very unlikely*. Often a hypothesis test is carried out at the 5% significance level.

The cut-off point c is known as the **critical value** and the set of observations that are considered to be unusual or unlikely (rare) events is called the **critical region** or **rejection region**. The critical value and critical region depend on the significance level chosen.

Suppose you choose a significance level of 5% to test Sid's claim.

From the working shown above, $P(X \geqslant 8) \approx 10\%$, which is greater than 5%, so 8 **is not** in the critical region. Getting 8 correct answers by guessing would not be considered an unlikely or rare event.

But $P(X \geqslant 9) \approx 4\%$, which is less than 5%, so any value of 9 or more **is** in the critical region. So getting 9 correct answers by guessing would be considered an unlikely or rare event.

To test Sid's claim at a **5% significance level**:

- the **critical value** is 9 correct answers
- the **critical region** is $X \geqslant 9$, i.e. 9, 10, 11, 12, ..., 19 or 20 correct answers.

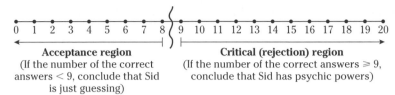

Acceptance region
(If the number of the correct answers < 9, conclude that Sid is just guessing)

Critical (rejection) region
(If the number of the correct answers $\geqslant 9$, conclude that Sid has psychic powers)

The language of hypothesis testing

The assumption that Sid is guessing is called the **null hypothesis** and is written H_0. The null hypothesis is very important as it provides the model for the calculations.

You would write

H_0: $p = 0.25$, where p is the probability that Sid identifies a card correctly

H_0 assumes Sid is guessing (with a 1 in 4 chance of guessing correctly)

If Sid has psychic powers, then he should get more than one in four correct and the probability that he gives the correct suit will be more than 0.25

This is called the **alternative hypothesis** and is denoted by H_1.

You would write

H_1: $p > 0.25$

H_1 assumes Sid has psychic powers

Having decided on the **significance level**, you can find the **critical (or rejection) region**. In this example, since you are interested in whether the probability is *greater* than 0.25, the critical region is in the *upper tail* (right-hand end) of the distribution and the test is known as a **one tail (upper tail) test**. Values that are not in the critical region are in the **acceptance region**.

The variable X, the number of correct answers, is the **test statistic**. Assuming that the null hypothesis is true, $X \sim B(20, 0.25)$

Note that the null hypothesis value of p is used in calculating probabilities.

The **test value** is the number of correct answers Sid gives in the experiment. To carry out the hypothesis test, you have to find whether or not the test value lies in the critical region. If it lies in the critical region it is considered an unusual event when p is 0.25

- If the test value lies in the **critical** region, you **reject H_0** in favour of H_1. This means that you reject the hypothesis that Sid is guessing, in favour of the alternative hypothesis that he is not guessing but has psychic powers.
- If the test value lies in the **acceptance** region, you do not have significant evidence to reject the null hypothesis and so you **accept H_0**, i.e. you do not have significant evidence to say that Sid has psychic powers so you accept that he is guessing.

Suppose Sid gives 7 correct answers, i.e. the test value is $x = 7$. Is this enough evidence, statistically, to say that he does have psychic powers?

From the critical region diagram above, you can see that, at the 5% level of significance, the test value of $x = 7$ does not lie in the critical region. Therefore you would not reject H_0. This means that there is not enough evidence that $p > 0.25$, i.e. to say that he has psychic powers. You would accept H_0, that Sid is guessing, and conclude that he does not have psychic powers.

Finding a probability

In practice the best way to find out whether the test value is in the critical region is to **calculate a probability**.

As Sid gave 7 correct answers, find $P(X \geqslant 7)$ and **compare** this probability with **5%** i.e. 0.05.

If $P(X \geqslant 7) < 0.05$, this means that 7 is in the critical region, so you would reject H_0.

Now $P(X \geqslant 7) = 1 - P(X \leqslant 6)$ Note: Calculate $P(X \geqslant 7)$, not $P(X = 7)$.

$$= 1 - (0.75^{20} + 20 \times 0.25 \times 0.75^{19} + \binom{20}{2} \times 0.25^2 \times 0.75^{18}$$

$$+ \binom{20}{3} \times 0.25^3 \times 0.75^{17} + \binom{20}{4} \times 0.25^4 \times 0.75^{16}$$

$$+ \binom{20}{5} \times 0.25^5 \times 0.75^{15} + \binom{20}{6} \times 0.25^6 \times 0.75^{14})$$

$$= 0.2142$$

Compare this with 5%:

Now $0.2142 > 0.05$, so 7 **is not** in the critical region. Therefore you would not reject H_0. You do not have enough evidence, at the 5% level, to say that Sid has psychic powers.

Now suppose that Sid gave 10 correct answers.

You will find that $P(X \geqslant 10) = 0.013\ldots$

As $0.013 < 0.05$, 10 **is** in the critical region. Therefore you would reject H_0 and conclude that it is very unlikely that Sid would identify 10 or more cards correctly just by guessing. Hence there is evidence at the 5% level to reject H_0. It is likely that Sid has psychic powers (or is cheating!).

Probability diagram

The diagram below shows the probability distribution for $X \sim B(20, 0.25)$, with the critical region indicated. Note that the distribution of X has a long tail to the right (a positive skew), with some probabilities being too small to show on the diagram.

You will see that 10 is in the critical region, whereas 7 is not in the critical region.

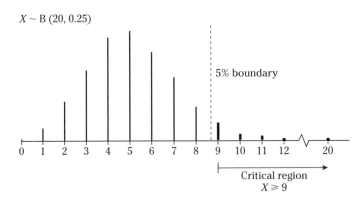

When, for example, you are testing $x = 7$ it may seem strange that you have to work out $P(X \geqslant 7)$ rather than just $P(X = 7)$. Remember that this is necessary as you are finding whether the test value lies in the critical (rejection) region or not.

Since $P(X \geqslant 8) \approx 10\%$ and $P(X \geqslant 9) \approx 4.1\%$, the 5% boundary comes between 8 and 9. The actual significance level is 4.1%. With discrete distributions you will not get a perfect 5% in your calculations.

Steps in a hypothesis test

When performing a hypothesis test, it is useful to work through the following steps. This will ensure that your argument is clear and that you do not leave out any important details.

- Define the variable
- State H_0 and H_1, then give the distribution of the test statistic assuming H_0 is true
- State the rejection rule – you could give a probability statement or define the critical region
- Find whether the test value lies in the critical region – this will usually involve finding a probability and comparing it with the significance level
- Make your conclusion in statistical terms by saying whether H_0 is rejected or not and then **relate it to the situation**

Example 5.1

A particular type of tomato seed has a germination rate of 72%.

In order to improve the germination rate, a manufacturer uses a new treatment, after which 14 seeds germinate in a pack of 15 seeds.

Stating suitable null and alternative hypotheses, test, at the 10% level, whether the germination rate has improved following the new treatment.

Follow the steps for a hypothesis test.

Define the variable	X is the number of seeds in 15 that germinate, where $X \sim B(15, p)$.
State H_0 and H_1 and the distribution of X	$H_0: p = 0.72$ (The germination rate is unchanged) $H_1: p > 0.72$ (The germination rate has increased) If H_0 is true, then $X \sim B(15, 0.72)$. $\quad p = 0.72, q = 1 - p = 0.28$
State the rejection rule	The test value is 14. Using a one-tail (upper tail) test, at the 10% significance level, 14 lies in the critical region if $P(X \geqslant 14) < 10\%$. So reject H_0 if $P(X \geqslant 14) < 0.1$.
Calculate $P(X \geqslant 14)$ and compare it with 10%	$P(X \geqslant 14) = P(X = 14) + P(X = 15)$ $\qquad = \binom{15}{14} \times 0.72^{14} \times 0.28 + 0.72^{15}$ $\qquad = 0.04950... < 0.1$ $\qquad$ So $x = 14$ lies in the critical region.
Make your conclusion, relating it to the situation.	Since $P(X \geqslant 14) < 0.1$, reject H_0 in favour of H_1. There is evidence, at the 10% level, that the germination rate has increased, supporting the manufacturer's claim.

Would the conclusion have been the same if only 13 seeds had germinated?

To check this, find $P(X \geqslant 13)$.

$P(X \geqslant 13) = P(X = 13) + P(X \geqslant 14)$

$\qquad = \binom{15}{13} \times 0.72^{13} \times 0.28^2 + 0.04950...$ (from above)

$\qquad = 0.1645... > 10\%$

So if 13 seeds germinate, H_0 is not rejected. There is not enough evidence, at the 10% level, to say that the germination rate has increased.

From the calculations on the previous page you can see that the critical region is $X \geqslant 14$. For your interest, the probability distribution showing the critical region is illustrated below. You would not, however, be expected to draw it as part of your solution.

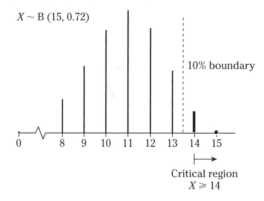

In the following example, the alternative hypothesis looks for a **decrease** in p, so the critical region is in the **lower tail**. This is known as a **one-tail (lower tail)** test.

Example 5.2

Isaac claims that 30% of cars in his town are red. His friend Hardip thinks that the proportion is less than 30%. The boys decided to test Isaac's claim at the 5% significance level and found that 2 cars out of a random sample of 18 were red. Carry out the hypothesis test and state your conclusion.

Cambridge Paper 7 Q1 N07

X is the number of red cars in a random sample of 18 cars, where $X \sim B(18, p)$.

H_0: $p = 0.3$ (The proportion of red cars is 30%, as Isaac claims)

H_1: $p < 0.3$ (The proportion of red cars is less than 30%, as Hardip claims)

If H_0 is true, then $X \sim B(18, 0.3)$.

*Since the alternative hypothesis is p < 0.3, the critical region is in the **lower tail** of the distribution.*

The test value is 2.

Using a one-tail (lower tail) test at the 5% significance level, 2 lies in the critical region if $P(X \leqslant 2) < 5\%$.

So reject H_0 if $P(X \leqslant 2) < 0.05$.

Now calculate the probability and compare it with 5%.

$P(X \leqslant 2) = P(X = 0) + P(X = 1) + P(X = 2)$

$\qquad = 0.7^{18} + 18 \times 0.3 \times 0.7^{17} + \binom{18}{2} \times 0.3^2 \times 0.7^{16}$

$\qquad = 0.0599\ldots > 0.05 \qquad$ This tells you that $x = 2$ is not in the critical region.

Since $P(X \leqslant 2) > 0.05$, do not reject H_0.

There is not enough evidence, at the 5% level, to reject Isaac's claim in favour of Hardip's claim. So you would accept Isaac's claim that the proportion of red cars is 30%.

Note: You cannot say with certainty that the proportion of red cars **is** 30%, but you do not have enough evidence to say that it is **less** than 30%.

Note: You will find that $P(X \leqslant 1) = 0.0141... < 5\%$, so the critical region is $X \leqslant 1$

This is shown on the sketch below.

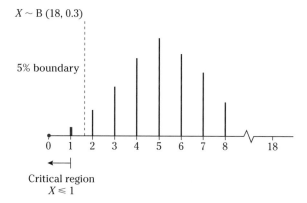

Finding the critical region

Example 5.3

In a game, Freddie wins points when a coin shows a head when it is tossed. After playing the game for some time, Freddie claims that the coin is biased in favour of tails. He decides to carry out a hypothesis test at the 5% level to determine whether the coin shows fewer heads than tails. He tosses the coin 20 times and counts the number of heads obtained.

   (i) Find the critical region for the test.

Freddie obtained 7 heads in 20 tosses of the coin.

   (ii) State, with a reason, whether there is evidence that the coin is biased in favour of tails.

   (i) X is the number of heads in 20 tosses of the coin where $X \sim B(20, p)$.

   H_0: $p = \frac{1}{2}$ (The coin is fair)

   H_1: $p < \frac{1}{2}$ (The coin is biased in favour of tails)

   If H_0 is true, then $X \sim B(20, 0.5)$

   *Since the alternative hypothesis is $p < \frac{1}{2}$, the critical region is in the **lower tail** of the distribution.*

   Use a one-tail (lower tail) test at the 5% significance level.

   Now find the critical region.

   You want the maximum value c such that $P(X \leqslant c) < 0.05$

   By guesswork, try $c = 5$

   $P(X \leqslant 5) = \left(\frac{1}{2}\right)^{20} + 20 \times \left(\frac{1}{2}\right)^{20} + \binom{20}{2} \times \left(\frac{1}{2}\right)^{20}$

   When calculating the probabilities, since $p = \frac{1}{2}$ and $q = \frac{1}{2}$, you have $\left(\frac{1}{2}\right)^{20}$ each time.

   $\qquad + \binom{20}{3} \times \left(\frac{1}{2}\right)^{20} + \binom{20}{4} \times \left(\frac{1}{2}\right)^{20} + \binom{20}{5} \times \left(\frac{1}{2}\right)^{20}$

   $\qquad = 0.0206... < 0.05$

   So the critical region includes 0, 1, 2, 3, 4, 5 heads.

But does it include 6 heads?

Try $c = 6$

$$P(X \leqslant 6) = P(X \leqslant 5) + P(X = 6)$$
$$= 0.0206... + \binom{20}{6} \times \left(\frac{1}{2}\right)^{20}$$
$$= 0.0576... > 0.05$$

So 6 is not in the critical region.

The critical region is $X \leqslant 5$, so Freddie will reject H_0 if he obtains 5 or fewer heads.

(ii) There were 7 heads in 20 tosses of the coin.

Since 7 does not lie in the critical region, Freddie would not reject H_0. There is not significant evidence that the coin is biased in favour of tails.

The probability distribution is shown below.

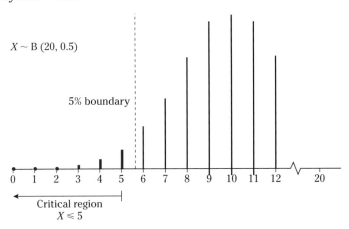

One-tail and two-tail tests

Consider a test with null hypothesis

$H_0: p = p_0$
p_0 is a specific value such as 0.3

One-tail tests

In the examples so far, one-tail tests have been considered, with either the upper or the lower tail being used for the critical region, depending on the alternative hypothesis.

A **one-tail test** is carried out when the alternative hypothesis looks for an **increase** or a **decrease** in p.

For a 5% significance level:

One tail (upper tail) test

If H_1 is $p > p_0$, then H_1 is looking for an **increase** in p.

The critical region will be in the upper tail and consists of values **greater than or equal to** c such that $P(X \geqslant c) < 5\%$, i.e. $P(X \geqslant c) < 0.05$

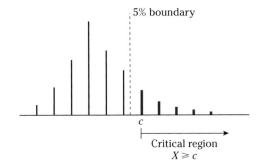

One-tail (lower tail) test

If H_1 is $p < p_0$, then H_1 is looking for a **decrease** in p.

The critical region will be in the **lower tail** and consists of values **less than or equal to c** such that
$P(X \leqslant c) < 5\%$, i.e. $P(X \leqslant c) < 0.05$

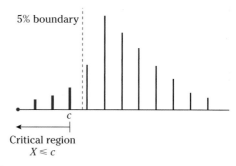

Remember that, in both cases, the probability **is less** than 0.05.

Two-tail tests

A **two-tail test** is carried out when the alternative hypothesis looks for a **change** in p, not specifically an increase or a decrease. The alternative hypothesis is that p does not equal p_0, i.e.

H_1: $p \neq p_0$

The critical region is in two parts, split between the upper tail and the lower tail.

For a 5% significance level:

In the **lower tail**, the critical region consists of values **less than or equal to c_1** such that

$P(X \leqslant c_1) < 2.5\%$, i.e. $P(X \leqslant c_1) < 0.025$

In the **upper tail**, the critical region consists of values **greater than or equal to c_2** such that

$P(X \geqslant c_2) < 2.5\%$, i.e. $P(X \geqslant c_2) < 0.025$

The critical region might look like this.

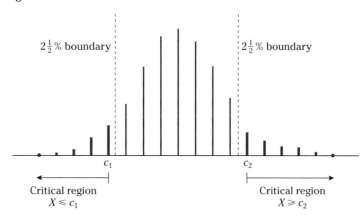

Summarising for a 5% significance level:

	Hypotheses	**Critical (rejection) region**	**Rejection rule** (x is the test value)
To test for an **increase** in p	H_0: $p = p_0$ H_1: $p > p_0$	Upper tail 5%	Reject H_0 if $P(X \geqslant x) < 5\%$
To test for a **decrease** in p	H_0: $p = p_0$ H_1: $p < p_0$	Lower tail 5%	Reject H_0 if $P(X \leqslant x) < 5\%$
To test for a **change** in p	H_0: $p = p_0$ H_1: $p \neq p_0$	Upper tail 2.5% Lower tail 2.5%	Reject H_0 if $P(X \geqslant x) < 2.5\%$ or $P(X \leqslant x) < 2.5\%$

Example 5.4

In last year's elections, the Purple party gained 35% of the votes. Before this year's election, a researcher was asked to test whether support for the Purple party had changed. She interviewed 12 voters, selected at random, and 1 voter said that he would vote for the Purple party.

(i) Test, at the 10% significance level, whether support for the Purple party has changed.

(ii) Find the critical region for the test.

(i) X is the number of voters in a randomly chosen group of 12 voters who say they will vote for the Purple party, where $X \sim B(12, p)$

H_0: $p = 0.35$ (The proportion is unchanged)

H_1: $p \neq 0.35$ (The proportion has changed)

If H_0 is true, then $X \sim B(12, 0.35)$

Use a two-tail test at the 10% level.

As this is a two-tail test, the 10% for the significance level is distributed between the lower and upper tails, with 5% in each tail.

Reject H_0 if $P(X \leq x) < 0.05$ or $P(X \geq x) < 0.05$, where x is the test value.

However, as the test value is 1, you need only look at the lower tail part of the critical region and find whether $P(X \leq 1) < 0.05$

$P(X \leq 1) = P(X = 0) + P(X = 1)$

$\qquad = 0.65^{12} + 12 \times 0.35 \times 0.65^{11}$

$\qquad = 0.0424... < 0.05$

Since $P(X \leq 1) < 0.05$, the test value 1 lies in the critical region, so H_0 is rejected in favour of H_1.

There is evidence, at the 10% level, that support for the Purple party has changed.

(ii) *To find the critical region, consider separately the upper and lower tails. You want 5% in each tail.*

<u>Critical region in lower tail:</u>

Find the greatest value c such that $P(X \leq c) < 0.05$

You already know from part (i) that $P(X \leq 1) < 0.05$, so the critical region contains 0 and 1

Try $c = 2$

$P(X \leq 2) = P(X = 0) + P(X = 1) + P(X = 2)$

$\qquad = 0.65^{12} + 12 \times 0.35 \times 0.65^{11} + \binom{12}{2} \times 0.35^2 \times 0.65^{10}$

$\qquad = 0.1512... > 0.05$

Since $P(X \leq 2) > 5\%$, $x = 2$ is not in the critical region.

The critical region in the lower tail is $X \leq 1$. So $x = 0, 1$ lie in the critical region.

<u>Critical region in upper tail:</u>

Find the least value c such that $P(X \geqslant c) < 0.05$

By guesswork, try $c = 8$

$P(X \geqslant 8) = P(X = 8) + P(X = 9) + P(X = 10) + P(X = 11) + P(X = 12)$

$$= \binom{12}{8} \times 0.35^8 \times 0.65^4 + \binom{12}{9} \times 0.35^9 \times 0.65^3$$

$$+ \binom{12}{10} \times 0.35^{10} \times 0.65^2 + \binom{12}{11} \times 0.35^{11} \times 0.65^1 + 0.35^{12}$$

$$= 0.0255... < 0.05$$

So the critical region includes 8, 9, 10, 11and 12.

But is 8 the smallest value in the critical region?

Try $c = 7$

$P(X \geqslant 7) = P(X = 7) + P(X \geqslant 8)$

$$= \binom{12}{7} \times 0.35^7 \times 0.65^5 + 0.0255.... \quad \text{(from above)}$$

$$= 0.084 ... > 0.05$$

So 7 is not in the critical region.

The critical region in the upper tail is $X \geqslant 8$, i.e. test values of 8, 9, 10, 11, 12 lie in the critical region.

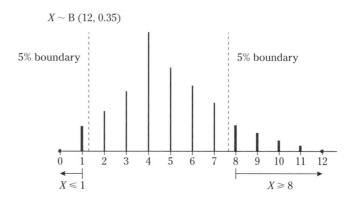

$X \sim B(12, 0.35)$

5% boundary

5% boundary

$X \leqslant 1$

$X \geqslant 8$

Therefore the critical (or rejection) region is $X \leqslant 1, X \geqslant 8$.

i.e. test values of $x = 0, 1, 8, 9, 10, 11, 12$ will lie in the critical region.

Exercise 5a

1 The records of a particular hospital show that 30% of patients who come to the hospital's casualty department have to wait more than an hour before receiving medical attention. An extra doctor is employed and during the next week 3 out of 20 randomly selected patients have to wait more than an hour to receive medical attention.

Carry out a hypothesis .est at the 5% level to test whether the proportion of patients who have to wait more than an hour has been reduced.

2 Rachel claims that 65% of students in her school have watched a particular film. Her friend Lily claims that more than 65% of students have watched the film. To test Rachel's claim, they decide to carry out a hypothesis test at the 10% significance level, asking 10 randomly selected students whether they have watched the film.

(i) State suitable null and alternative hypotheses.

(ii) Find the critical region for the test.

In the random sample of 10 students, 9 had watched the film.

(iii) State the conclusion of the test.

(iv) Would your conclusion be different if Lily had claimed that the proportion of students who have watched the film was not 65%?

3 The manager at a particular supermarket claims that fewer than 35% of people pay by cheque. In a random sample of 16 people, 3 people pay by cheque. Test, at the 10% significance level, the null hypothesis $H_0: p = 0.35$ against the alternative hypothesis $H_1: p < 0.35$. State your conclusion clearly.

4 Hester suspected that a die was biased in favour of a four occurring. She decided to carry out a hypothesis test.

(i) State suitable null and alternative hypotheses for the test.

When she threw the die 15 times, she obtained a four on 6 occasions.

(ii) Carry out the test, at the 5% level, stating your conclusion clearly.

5 A pharmaceutical company produced a new pain-relieving drug for migraine sufferers and its advertisements stated that the drug has a 90% success rate. A doctor doubted whether the drug would be as successful as the company claimed. She prescribed the drug for 15 of her patients. After 6 months, 11 of these patients said that their migraine symptoms had been relieved by the drug.

Test the pharmaceutical company's claim, at the 5% significance level.

6 The random variable X can be modelled by a binomial distribution with $n = 10$

A random observation, x, is taken from the distribution.

Test, at the 8% level, the hypothesis that $p = 0.44$ against the alternative hypothesis $p \neq 0.44$:

(i) when $x = 7$

(ii) when $x = 1$.

7 In each of the following, a random observation x is taken from a binomial distribution $X \sim B(n, p)$

Test the given hypotheses at the significance level stated.

	Sample value x	n	Hypotheses	Significance level
(i)	1	10	$H_0: p = 0.45$ $H_1: p < 0.45$	5%
(ii)	6	8	$H_0: p = 0.45$ $H_1: p > 0.45$	5%
(iii)	5	7	$H_0: p = 0.25$ $H_1: p \neq 0.25$	2%
(iv)	2	20	$H_0: p = 0.3$ $H_1: p < 0.3$	1%
(v)	12	14	$H_0: p = 0.58$ $H_1: p \neq 0.58$	6%
(vi)	8	12	$H_0: p = 0.8$ $H_1: p < 0.8$	10%

8 In a test of 10 true-false questions, Sian got 8 correct. Carry out a hypothesis test at the 5% significance level to test whether she could have obtained this score by guessing all the answers.

9 A survey is being carried out in a particular city on the methods of transport used by commuters to travel to work.

The local authority says that 40% of commuters cycle to work, but the Cycling Association claims that more than 40% of commuters cycle to work. In a random sample of 12 commuters, 8 cycle to work.

Carry out a hypothesis test at the 6% significance level to test the Cycling Association's claim.

10 When Alex used to play darts regularly he scored a bull on 40% of attempts, on average. After a break of three months, he played darts one evening and scored 2 bulls on 12 attempts.

Stating a necessary assumption, carry out a hypothesis test at the 10% significance level, to test whether the percentage of bulls he scores has decreased.

11 The random variable X has distribution B(10, p). A hypothesis test, at the 10% significance level, is to be carried out to test the null hypothesis $p = 0.5$ against the alternative hypothesis $p \neq 0.5$

Find the critical region for the test.

12 A large college introduced a new procedure to try to ensure that staff arrived on time for the start of lectures. A recent survey by the students had suggested that, in 15% of cases, staff arrived late for the start of the lecture. In the first week following the introduction of the new procedure a random sample of 35 lectures was taken and in only one case did the member of staff arrive late.

(i) Stating your hypotheses clearly, test, at the 5% level of significance, whether there is evidence that the new procedure has been successful.

A student complained that this sample did not give a true indication of the effectiveness of the new procedure.

(ii) Explain how the student's claim might be justified and suggest how a more effective check on the new procedure could be made.

HYPOTHESIS TEST FOR A POISSON MEAN

The procedure for carrying out a hypothesis test for the mean λ of a Poisson distribution is the same as that for the binomial proportion.

The null hypothesis is

$H_0: \lambda = \lambda_0$ λ_0 is a specific value.

The alternative hypothesis H_1 depends on whether the test is one-tail or two-tail.

For a 5% significance level:

	Hypotheses	**Critical (rejection) region**	**Rejection rule** (x is the test value)
To test for an **increase** in λ	$H_0: \lambda = \lambda_0$ $H_1: \lambda > \lambda_0$	Upper tail 5%	Reject H_0 if $P(X \geqslant x) < 5\%$
To test for a **decrease** in λ	$H_0: \lambda = \lambda_0$ $H_1: \lambda < \lambda_0$	Lower tail 5%	Reject H_0 if $P(X \leqslant x) < 5\%$
To test for a change in λ	$H_0: \lambda = \lambda_0$ $H_1: \lambda \neq \lambda_0$	Upper tail 2.5% Lower tail 2.5%	Reject H_0 if $P(X \geqslant x) < 2.5\%$ or $P(X \leqslant x) < 2.5\%$

Example 5.5

The number of misprints on the advertisements page of the Daily Informer has a Poisson distribution with mean 6.5 misprints per page. When a new typesetter is employed, 3 misprints are found on a randomly selected advertisements page. The new typesetter claims that the mean number of misprints has decreased. Test this claim at the 5% significance level.

X is the number of misprints on the advertisements page, where $X \sim \text{Po}(\lambda)$.

$H_0: \lambda = 6.5$ (The mean number of misprints is unchanged)

$H_1: \lambda < 6.5$ (The mean number of misprints has decreased)

If H_0 is true, then $X \sim \text{Po}(6.5)$.

*Since the alternative hypothesis is $\lambda < 6.5$, the critical region is in the **lower tail** of the distribution.*

The test value is 3.

Using a one-tail (lower tail) test at the 5% significance level, 3 lies in the critical region if $P(X \leqslant 3) < 5\%$.

So reject H_0 if $P(X \leqslant 3) < 0.05$.

Now calculate the probability and compare it with 5%.

$$P(X \leqslant 3) = e^{-6.5}\left(1 + 6.5 + \frac{6.5^2}{2!} + \frac{6.5^3}{3!}\right)$$

$$= 0.1118... > 0.05$$

Recall: Poisson probabilities (page 2)

So 3 is not in the critical region.

Since $P(X \leqslant 3) > 0.05$, do not reject H_0.

There is not enough evidence, at the 5% level, that the mean number of misprints has decreased, so the new typesetter's claim in not supported.

Example 5.6

Jack is the manager of the local football team. During last season the number of goals per match scored by the team followed a Poisson distribution with mean 1.4

Prior to the start of the new season, the team took part in intensive coaching sessions. Jack decided to use the results of the first 2 matches of the new season to test whether the team now scores more goals than before, on average.

In the first 2 matches of the new season, the total number of goals scored by the team was 6.

 (i) Carry out the test at the $7\frac{1}{2}\%$ significance level, clearly stating your hypotheses and conclusion.

 (ii) State, with a reason, whether the conclusion would be the same at the 5% significance level.

X is the number of goals in **2 matches**, where $X \sim Po(\lambda)$.

Note: λ is the mean number of goals in **2 matches**.

H_0: $\lambda = 2.8$ (The mean number of goals is unchanged)

H_1: $\lambda > 2.8$ (The mean number of goals has increased)

You need to consider 2 matches because the sample data gives the results of 2 matches.

$\lambda = 2 \times 1.4 = 2.8$

If H_0 is true, then $X \sim Po(2.8)$.

 (i) The test value is 6.

   Using a one-tail (upper tail) test at the $7\frac{1}{2}\%$ significance level, reject H_0 if $P(X \geqslant 6) < 0.075$.

   $$P(X \geqslant 6) = 1 - P(X \leqslant 5)$$

   $$= 1 - e^{-2.8}\left(1 + 2.8 + \frac{2.8^2}{2!} + \frac{2.8^3}{3!} + \frac{2.8^4}{4!} + \frac{2.8^5}{5!}\right)$$

   $$= 1 - 0.9348...$$

   $$= 0.06511... < 0.075$$

   This tells you that $x = 6$ is in the critical region.

   Since $P(X \geqslant 6) < 0.075$, reject H_0.

   There is evidence, at the $7\frac{1}{2}\%$ level, that the mean number of goals per match has increased.

(ii) At the 5% significance level, you would reject H_0 if $P(X \geqslant 6) < 0.05$

From part (i), $P(X \geqslant 6) = 0.06511... > 0.05$

> At the **5% level**, $x = 6$ is **not** in the critical region.

So, at the 5% level, you would not reject H_0. There is not enough evidence, at the 5% level, to support the claim that the mean number of goals per match has increased.

Exercise 5b

1   The number of minor accidents at a particular factory may be modelled by a Poisson distribution with mean 4.1 per week. Health and safety advice is given to all employees, following which the number of accidents in a randomly selected week was 2. Does this provide evidence, at the 5% significance level, that the mean number of accidents per week has decreased?

2   Over a period of time it is found that the number of reported cases of a particular medical condition follows a Poisson distribution with mean 1.4 per month.

During a particular month, 4 cases are reported. Stating a necessary assumption, perform a hypothesis test at the 5% level to test whether the mean number of reported cases per month of this medical condition is greater than 1.4

3   Over a period of time it is found that the number of letters of complaint received by a large store follows a Poisson distribution with mean 1.1 per day. Following the appointment of a new manager, it is believed that the mean number of letters of complaint has changed. In a randomly selected period of 5 days, 2 letters of complaint were received.

Carry out a hypothesis test, at the 10% level, to test whether the mean number of letters of complaint has changed.

4   The number of telephone calls made to an office may be modelled by a Poisson distribution with mean 6 per hour.

(i) In a particular 30 minute period, 5 calls were received. Test, at the 5% level, whether the mean number of calls has increased.

(ii) On a particular day, there were 3 calls between 11.00am and 12.30pm. Test, at the 5% level whether the mean number of calls has decreased.

5   In each of the following, a random observation x is taken from a Poisson distribution X, where $X \sim Po(\lambda)$

Test the given hypotheses at the significance level stated.

	Sample value x	Hypotheses	Significance level
(i)	4	$H_0: \lambda = 10$ $H_1: \lambda < 10$	5%
(ii)	2	$H_0: \lambda = 8.1$ $H_1: \lambda \neq 8.1$	10%
(iii)	3	$H_0: \lambda = 7$ $H_1: \lambda \neq 7$	8%
(iv)	4	$H_0: \lambda = 1.4$ $H_1: \lambda > 1.4$	3%

6   The number of breakdowns per day of the lifts in a large block of flats has a Poisson distribution with mean 0.2

The maintenance contract for the lists is given to a new company. With this company it is found that there are 2 breakdowns over a period of 30 days. Carry out a hypothesis test at the 5% level to decide whether the mean number of breakdowns has decreased.

7   The random variable X is distributed $Po(\lambda)$. A hypothesis test at the 5% level is to be carried out to test the null hypothesis $\lambda = 9$ against the alternative hypothesis $\lambda < 9$. The critical region is $X \leqslant c$. Find the value of c.

8 The number of bacterial colonies that develop in dishes of nutrient exposed to an infected environment has a Poisson distribution with mean 7.5

An experiment was conducted to determine the effectiveness of an antibiotic spray. When a dish was sprayed, the number of bacterial colonies that developed was 3. Stating suitable null and alternative hypotheses, test, at the 5% significance level, whether the mean number of bacterial colonies has decreased.

9 The number of customers arriving per half-hour at a shop has a Poisson distribution with mean 2.2. After a sales campaign, it is hoped that the mean will increase. The manager decides to note how many customers arrive during a randomly chosen half-hour period the next day and to use this figure to carry out a hypothesis test at the 5% level. Calculate the critical region.

TYPE I AND TYPE II ERRORS

When you carry out a hypothesis test there are four possible conclusions, two of which lead to a correct decision being made and the other two to a wrong decision being made. The **errors** associated with making wrong decisions are called **Type I** and **Type II** errors.

The outcomes and errors are summarised as follows:

- H_0 is true and your test leads you to accept H_0: correct decision
- H_0 is true but your test leads you to reject H_0: wrong decision, Type I error
- H_0 is false but your test leads you to accept H_0: wrong decision, Type II error
- H_0 is false and your test leads you to reject H_0: correct decision.

Type I errors

A **Type I error** is made when you reject H_0 when in fact H_0 is true.

P(Type I error) = P(reject H_0 when H_0 is true)

Example 5.7

A manufacturer claims that 20% of sugar-coated chocolate beans are red. George suspects that this percentage is actually less than 20% and so he takes a random sample of 15 chocolate beans and performs a hypothesis test with the null hypothesis $p = 0.2$ against the alternative hypothesis $p < 0.2$. He decides to reject the null hypothesis in favour of the alternative hypothesis if there are 0 or 1 red beans in the sample.

(i) With reference to this situation, explain what is meant by a Type I error.

(ii) Find the probability of a Type I error in George's test.

Cambridge Paper 7 Q2 N05

(i) George will make a Type I error if he concludes that the percentage of red beans is less than 20% when, in fact, it is 20%

(ii) X is the number of red beans in a random sample of 15 chocolate beans, where $X \sim B(15, p)$.

H_0: $p = 0.2$ (20% are red, as claimed by the manufacturer)

H_1: $p < 0.2$ (less than 20% are red, as George suspects)

If H_0 is true, then $X \sim B(15, 0.2)$

The rejection rule is that George will reject H_0 if $X \leqslant 1$

P(Type I error)

    = P(reject H_0 when H_0 is true)

    = P($X \leqslant 1$ when $X \sim$ B(15, 0.2))

    = P($X = 0$) + P($X = 1$)

    = $0.8^{15} + 15 \times 0.2 \times 0.8^{14}$

    = $0.1671...$

So P(Type I error) $\approx 16.7\%$

Example 5.8

A single random observation is taken from a Poisson distribution with mean λ and used to test, at the 5% significance level, the null hypothesis $\lambda = 8$ against the alternative hypothesis $\lambda < 8$. Find the critical region and hence find the probability of making a Type I error.

(i) $X \sim$ Po(λ).

    $H_0: \lambda = 8$

    $H_1: \lambda < 8$

    If H_0 is true, then $X \sim$ Po(8).

    To find the critical region, you want the greatest value c such that P($X \leqslant c$) < 0.05.

    Try $c = 3$
$$P(X \leqslant 3) = e^{-8}\left(1 + 8 + \frac{8^2}{2!} + \frac{8^3}{3!}\right)$$
$$= 0.04238... < 0.05$$

    So 0, 1, 2 and 3 lie in the critical region.

    Now try $c = 4$
$$P(X \leqslant 4) = P(X \leqslant 3) + P(X = 4)$$
$$= 0.0423... + e^{-8} \times \frac{8^4}{4!}$$
$$= 0.0996... > 0.05$$

    So 4 does not lie in the critical region.

    The critical region is $X \leqslant 3$.

    *You make a Type I error when you reject H_0 when it is in fact true, i.e. you reject the hypothesis that the mean is 8 when in fact it **is** 8.*

    Since the critical region is $X \leqslant 3$, you would reject H_0 when $X \leqslant 3$.

    P(Type I error)

        = P(reject H_0 when H_0 is true)

        = P($X \leqslant 3$ when $X \sim$ Po(8))

        = $0.04238...$ (from above)

        = 0.0424 (3 s.f.)

    So P(Type I error) $\approx 4.24\%$

Type II errors

A **Type II error** is made when you accept H_0 when in fact H_0 is false.

To calculate the probability of making a Type II error, you **must be given a specific value for the alternative hypothesis H_1**.

Then $\quad$ P(Type II error) = P(accept H_0 when H_1 is true)

Example 5.9

The number of severe floods per year in a certain country over the last 100 years has followed a Poisson distribution with mean 1.8. Scientists suspect that global warming has now increased the mean. A hypothesis test, at the 5% level, is to be carried out to test this suspicion. The number of severe floods, X, that occur next year will be used for the test.

(i) Show that the rejection region for the test is $X > 4$.

(ii) Find the probability of making a Type II error if the mean number of severe floods is now actually 2.3. $\hfill$ Cambridge Paper 7 Q4 N09(71)

(i) X is the number of severe floods in a year, where $X \sim \text{Po}(\lambda)$

$H_0: \lambda = 1.8$ (The mean number of severe floods is unchanged)

$H_1: \lambda > 1.8$ (The mean number of severe floods has increased)

If H_0 is true, then $X \sim \text{Po}(1.8)$

*Since the alternative hypothesis is $\lambda > 1.8$, the rejection (critical) region is in the **upper tail** of the distribution.*

To find the rejection region, first consider $P(X > 4)$

$$P(X > 4) = 1 - P(X \leqslant 4)$$
$$= 1 - e^{-1.8}\left(1 + 1.8 + \frac{1.8^2}{2!} + \frac{1.8^3}{3!} + \frac{1.8^4}{4!}\right)$$
$$= 1 - 0.9635\ldots$$
$$= 0.03640\ldots < 0.05$$

So values of X greater than 4 are in the rejection region, i.e. 5, 6, 7, … are in the rejection region.

But is 4 in the rejection region as well?

$$P(X \geqslant 4) = P(X = 4) + P(X > 4)$$
$$= e^{-1.8} \times \frac{1.8^4}{4!} + 0.03640\ldots$$
$$= 0.1087\ldots > 0.05$$

Since $P(X \geqslant 4) > 0.05$, 4 is not in the rejection region.

Therefore the rejection region is $X > 4$.

(ii) You reject H_0 if $X > 4$, so you accept H_0 if $X \leqslant 4$

A Type II error is made when you accept H_0 when in fact it is not true.

If the true value of λ is 2.3, then

P(Type II error)
$$= P(X \leqslant 4 \text{ when } X \sim \text{Po}(2.3))$$
$$= e^{-2.3}\left(1 + 2.3 + \frac{2.3^2}{2!} + \frac{2.3^3}{3!} + \frac{2.3^4}{4!}\right)$$
$$= 0.9162\ldots$$
$$= 0.916 \text{ (3 s.f.)}$$

Example 5.10

Joanna thinks that a box of counters contains 10 white counters and 90 black counters. Kate thinks that the box contains 50 white counters and 50 black counters. In order to test these assumptions they decide to take 4 counters at random from the box, replacing each counter in the box after its colour has been noted, and to accept Joanna's hypothesis (the null hypothesis) if all 4 counters are black.

(i) State what is meant by a Type I error in the context of this question, and find the probability that the test results in a Type I error.

(ii) If Kate's hypothesis (the alternative hypothesis) is true, calculate the probability of a Type II error.

H_0: The box contains 10 white counters and 90 black counters
        (Joanna's claim)

H_1: The box contains 50 white counters and 50 black counters
        (Kate's claim)

(i) H_0 is accepted if all 4 counters are black, so H_0 is rejected if at least 1 of the 4 counters drawn is white.

P(Type I error)

$= $ P(reject H_0 when H_0 is true)

$= $ P(at least 1 white counter when the box contains 10 white and 90 black counters)

$= 1 - $ P(all black) $P(\text{black}) = \frac{90}{100}$

$= 1 - \left(\frac{90}{100}\right)^4$

$= 0.3439 \approx 34\%$

(ii) P(Type II error)

$= $ P(accept H_0 when H_1 is true)

$= $ P(all 4 are black when the box contains 50 white and 50 black counters)

$= \left(\frac{50}{100}\right)^4$ $P(\text{black}) = \frac{50}{100}$

$= 0.0625 \approx 6.25\%$

Example 5.11

Kevin has a cubical die which he suspects is biased so that a six occurs more often than expected if the die is fair. He decides to carry out a hypothesis test. The null hypothesis is $p = \frac{1}{6}$, where p is the probability of throwing a six, and the alternative hypothesis is $p > \frac{1}{6}$. He decides to reject the null hypothesis in favour of the alternative hypothesis if he obtains 4 or more sixes when he throws the die 12 times.

(i) Find the probability of making a Type I error.

(ii) Find the probability of making a Type II error if the probability of throwing a six is $\frac{1}{3}$.

X is the number of sixes when the die is thrown 12 times, where $X \sim \mathrm{B}(12, p)$

(i) H_0: $p = \frac{1}{6}$ (the die is fair)

H_1: $p > \frac{1}{6}$ (the die is biased in favour of sixes)

If H_0 is true, then $X \sim \mathrm{B}(12, \frac{1}{6})$

The rejection rule is that Kevin will reject H_0 if $X \geqslant 4$

P(Type I error)

$$= \mathrm{P}(\text{reject } \mathrm{H}_0 \text{ when } \mathrm{H}_0 \text{ is true})$$

$$= \mathrm{P}(X \geqslant 4 \text{ when } X \sim \mathrm{B}(12, \tfrac{1}{6}))$$

$$= 1 - \mathrm{P}(X \leqslant 3)$$

$$= 1 - (\mathrm{P}(X = 0) + \mathrm{P}(X = 1) + \mathrm{P}(X = 2) + \mathrm{P}(X = 3))$$

$$= 1 - \left[\left(\tfrac{5}{6}\right)^{12} + 12 \times \left(\tfrac{1}{6}\right) \times \left(\tfrac{5}{6}\right)^{11} + \binom{12}{2} \times \left(\tfrac{1}{6}\right)^2 \times \left(\tfrac{5}{6}\right)^{10} + \binom{12}{3} \times \left(\tfrac{1}{6}\right)^3 \times \left(\tfrac{5}{6}\right)^9 \right]$$

$$= 0.1251...$$

So P(Type I error) $\approx 12.5\%$

(ii) H_0 is accepted if $X < 4$, i.e. $X \leqslant 3$

P(Type II error)

$$= \mathrm{P}(\text{accept } \mathrm{H}_0 \text{ when } \mathrm{H}_1 \text{ is true})$$

$$= \mathrm{P}(X \leqslant 3 \text{ when } X \sim \mathrm{B}(12, \tfrac{1}{3}))$$

$$= \mathrm{P}(X = 0) + \mathrm{P}(X = 1) + \mathrm{P}(X = 2) + \mathrm{P}(X = 3)$$

$$= \left(\tfrac{2}{3}\right)^{12} + 12 \times \left(\tfrac{1}{3}\right) \times \left(\tfrac{2}{3}\right)^{11} + \binom{12}{2} \times \left(\tfrac{1}{3}\right)^2 \times \left(\tfrac{2}{3}\right)^{10} + \binom{12}{3} \times \left(\tfrac{1}{3}\right)^3 \times \left(\tfrac{2}{3}\right)^9$$

$$= 0.3930...$$

So P(Type II error) $\approx 39.3\%$

Example 5.12

In a certain city it is necessary to pass a driving test in order to be allowed to drive a car. The probability of passing the driving test at the first attempt is 0.36 on average. A particular driving instructor claims that the probability of his pupils passing at the first attempt is higher than 0.36. A random sample of 8 of his pupils showed that 7 passed at the first attempt.

(i) Carry out an appropriate hypothesis test to test the driving instructor's claim, using a significance level of 5%.

(ii) In fact, most of this random sample happened to be careful and sensible drivers. State which type of error in the hypothesis test (Type I or Type II) could have been made in these circumstances and find the probability of this type of error when a sample of size 8 is used for the test.

Cambridge Paper 7 Q4 J09(71)

(i) X is the number of pupils in 8 who pass the driving test at the first attempt, where $X \sim B(8, p)$.

H_0: $p = 0.36$ (The proportion is unchanged)

H_1: $p > 0.36$ (The proportion is greater than 36%)

If H_0 is true, then $X \sim B(8, 0.36)$.

The test value is 7.

Using a one-tail (upper tail) test at the 5% significance level, reject H_0 if $P(X \geqslant 7) < 0.05$.

$P(X \geqslant 7) = P(X = 7) + P(X = 8)$

$$= \binom{8}{7} \times 0.36^7 \times 0.64 + 0.36^8$$

$$= 0.004294... < 0.05 \qquad \text{This indicates that } x = 7 \text{ is in the critical region.}$$

Since $P(X \geqslant 7) < 0.05$, H_0 is rejected in favour of H_1.

There is evidence, at the 5% level, that the probability that the driving instructor's pupils pass at the first attempt is higher than 0.36, so his claim is supported.

(ii) H_0 is rejected, so a Type I error could have been made.

To find the probability of a Type I error, find the critical region.

You want to find the least value c such that P(X ⩾ c) < 0.05.

From part (i) you know that the critical region contains 7 and 8.

Try $c = 6$.

$P(X \geqslant 6) = P(X = 6) + P(X \geqslant 7)$

$$= \binom{8}{6} \times 0.36^6 \times 0.64^2 + 0.004294...$$

$$= 0.02925... < 0.05$$

So the critical region involudes 6, 7 and 8

Now try $c = 5$

$P(X \geqslant 5) = P(X = 5) + P(X \geqslant 6)$

$$= \binom{8}{5} \times 0.36^5 \times 0.64^3 + 0.02925...$$

$$= 0.1189... > 0.05$$

So 5 is not in the critical region.

Therefore the critical region is $X \geqslant 6$.

So, if 6 or more pupils pass at the first attempt, you would reject H_0, even though it is true.

P(Type I error) $= P(X \geqslant 6) = 0.02925... = 0.0293$ (3 s.f.)

Example 5.13

A hospital patient's white blood cell count has a Poisson distribution. Before undergoing treatment the patient had a mean white blood cell count of 5.2. After the treatment a random measurement of the patient's white blood cell count is made, and is used to test at the 10% significance level whether the mean white blood cell count has decreased.

(i) State what is meant by a Type I error in the context of the question, and find the probability that the test results in a Type I error.

(ii) Given that the measured value of the white blood cell count after the treatment is 2, carry out the test.

(iii) Find the probability of a Type II error if the mean white blood cell count after treatment is actually 4.1

Cambridge Paper 7 Q7 J10(71)

(i) A Type I error is made when you reject H_0 when it is in fact true, i.e. when you say that the white blood cell count has decreased when it has not decreased.

X is the white blood count, where $X \sim Po(\lambda)$

H_0: $\lambda = 5.2$ (The mean white blood cell count is unchanged)

H_1: $\lambda < 5.2$ (The mean white blood cell count has decreased)

If H_0 is true, then $X \sim Po(5.2)$

Use a one-tail (lower tail) test at the 10% significance level.

To find the probability of a Type I error, you need to find the rejection (critical) region.

$P(X = 0) = e^{-5.2} = 0.00516... < 10\%$

$P(X \leq 1) = P(X = 0) + P(X = 1)$

$\qquad = e^{-5.2}(1 + 5.2)$

$\qquad = 0.03420... < 10\%$

$P(X \leq 2) = P(X = 0) + P(X = 1) + P(X = 2)$

$\qquad = e^{-5.2}\left(1 + 5.2 + \dfrac{5.2^2}{2!}\right)$

$\qquad = 0.1087... > 10\%$

Since $P(X \leq 1) < 10\%$ and $P(X \leq 2) > 10\%$, the rejection region is $X \leq 1$. So sample values of 0 and 1 lie in the rejection region.

Now you can find the probability of a Type I error.

You would reject H_0, even when it is true, when $x = 0$ or 1

So, P(Type I error) = $P(X \leq 1$ when $X \sim Po(5.2))$

> Although the nominal significance level of the test is 10%, the **actual** significance level is 3.42%

$\qquad = 0.03420...$

$\qquad = 3.42\%$ (3 s.f.)

(ii) The test value is 2. This is not in the rejection region, so H_0 is not rejected.

There is not enough evidence of a decrease in the mean white blood cell count after treatment.

(iii) A Type II error is made when you accept H_0 when it is not true.

The acceptance region for H_0 is $X \geqslant 2$

So, if the true value of λ is 4.1,

P(Type II error)

$$= P(X \geqslant 2 \text{ when } X \sim Po(4.1))$$

$$= 1 - (P(X = 0) + P(X = 1))$$

$$= 1 - e^{-4.1}(1 + 4.1)$$

$$= 0.9154...$$

$$= 0.915 \text{ (3 s.f.)}$$

So P(Type II error) $\approx 91.5\%$

Example 5.14

In the past, the number of house sales completed per week by a building company has been modelled by a random variable which has the distribution Po(0.8). Following a publicity campaign, the builders hope that the mean number of sales per week will increase. In order to test at the 5% significance level whether this is the case, the total number of sales during the first three weeks after the campaign is noted. It is assumed that a Poisson model is still appropriate.

(i)   Given that the total number of sales during the 3 weeks is 5, carry out the test.

(ii) During the following 3 weeks the same test is carried out again, using the same significance level. Find the probability of a Type I error.

(iii) Explain what is meant by a Type I error in this context.

(iv) State what further information would be required in order to find the probability of a Type II error.

Cambridge Paper 7 Q7 N10(73)

(i)   Before the publicity campaign, the mean number of sales per week is 0.8, so the mean number of sales in 3 weeks is 2.4.

> You need to consider a 3-week period as this is the interval for the sample data.

X is the number of sales in 3 weeks, where $X \sim Po(\lambda)$.

[λ is the mean number of sales in 3 weeks.]

$H_0: \lambda = 2.4$ (The mean number of sales is unchanged)

$H_1: \lambda > 2.4$ (The mean number of sales has increased)

If H_0 is true, then $X \sim Po(2.4)$

There were 5 sales in the 3-week period, so the test value is 5.

Using a one-tail (upper tail) test at the 5% significance level, reject H_0 if $P(X \geqslant 5) < 0.05$.

$$P(X \geqslant 5) = 1 - P(X < 5)$$

$$= 1 - e^{-2.4}\left(1 + 2.4 + \frac{2.4^2}{2!} + \frac{2.4^3}{3!} + \frac{2.4^4}{4!}\right)$$

$$= 1 - 0.9041...$$

$$= 0.0958... > 0.05$$

> $x = 5$ in **not** in the critical region.

Since $P(X \geq 5) > 0.05$, do not reject H_0.

There is not enough evidence, at the 5% level, that the mean number of sales has increased following the publicity campaign.

(ii) To find P(Type I error) you need to find the least value c such that $P(X \geq c) < 0.05$

From part (i) you know that $P(X \geq 5) > 0.05$, so you know that c is greater than 5.

Try $c = 6$

$P(X \geq 6)$

$$= 1 - P(X < 6)$$
$$= 1 - e^{-2.4}\left(1 + 2.4 + \frac{2.4^2}{2!} + \frac{2.4^3}{3!} + \frac{2.4^4}{4!} + \frac{2.4^5}{5!}\right)$$
$$= 1 - 0.96431...$$
$$= 0.03567... < 0.05$$

So the least value is 6, and the critical region is $X \geq 6$, i.e. you would reject H_0 if the test value is 6, 7, 8, ...

So P(Type I error)

$$= P(\text{reject } H_0 \text{ when } H_0 \text{ is true})$$
$$= P(X \geq 6 \text{ when } X \sim Po(\,2.4))$$
$$= 0.03567...$$
$$= 0.0357 \text{ (3 s.f.)}$$

(iii) The mean number of sales in 3 weeks is still 2.4, but if the total number of sales is 6 or more you would conclude that the mean number of sales has increased, even though it has not.

(iv) A Type II error is made when you accept H_0 when H_0 is false. To find the probability of making a Type II error you would need to know the actual figure for the mean sales per week after the publicity.

Exercise 5c

1 The random variable X is distributed $B(9, p)$ where p is unknown.

(i) A hypothesis test is carried out at the 10% level to test the null hypothesis $p = 0.45$ against the alternative hypothesis $p < 0.45$. Find the critical region.

(ii) Explain what is meant by a Type I error.

(iii) Find the probability of making a Type I error in the test described in part (i).

2 At a particular hospital it was found from past records that the probability that a patient does not turn up for an appointment is 0.3

Following a campaign to make patients more aware of the problems caused by missed appointments, a significance test at the 10% level was carried out to decide whether the campaign had been effective in reducing the number of patients who did not turn up for an appointment. A random sample of 16 patients was surveyed.

(i) Find the critical region for the test.

(ii) Find the probability of making a Type II error in the test described in part (i) if the probability that a patient does not turn up for an appointment is, in fact, 0.25

3   Jessica thinks that a particular coin is biased in favour of heads. She decides to carry out a hypothesis test. The null hypothesis is $p = \frac{1}{2}$, where p is the probability of obtaining a head. The alternative hypothesis is $p > \frac{1}{2}$

She decides to conclude that the coin is biased in favour of heads if more than two-thirds of the tosses result in heads when she tosses the coin 12 times.

   (i)   (a)   Explain what is meant by making a Type I error in this context.

        (b)   Find the probability of making a Type I error.

   (ii) Find the probability of making a Type II error if the probability of throwing a head is, in fact, 0.7

4   A random observation is taken from the distribution of X where $X \sim B(20, p)$

A hypothesis test is carried out. The null hypothesis is $p = 0.4$ and the alternative hypothesis is $p < 0.4$

The critical region is chosen to be $X \leqslant 4$

   (i)   Find the probability of making a Type I error.

   (ii) Find the probability of making a Type II error if $p = 0.3$

5   A single observation is to be taken from a Poisson distribution, where $X \sim Po(\lambda)$, and used to test the null hypothesis $\lambda = 6$ against the alternative hypothesis $\lambda < 6$

The critical region is chosen to be $X \leqslant 2$

   (i)   Find the probability of making a Type I error.

   (ii) Find the probability of making a Type II error if $\lambda = 3.6$

6   The number of breakdowns per month of an office coffee machine may be modelled by a Poisson distribution. Before it was serviced, the mean number of breakdowns per month was 1.8

In the four months following the service, the machine broke down 3 times.

   (i)   Test, at the 5% level of significance, whether there has been an improvement in the reliability of the coffee machine.

Later a similar test at the 5% significance level was carried out using the number of breakdowns from another four months.

   (ii) Find the probability of making a Type I error.

7   A manufacturer of windows has used a process which produced flaws in the glass randomly at a rate of 0.5 per m^2. In an attempt to reduce the number of flaws produced, a new process is tried out. A randomly chosen window produced using this new process has an area of 8 m^2 and contains 2 flaws.

   (i)   Stating your hypotheses clearly, test at the 10% significance level whether the rate of occurrence of flaws using the new procedure has decreased.

The new procedure actually produces flaws at a rate of 0.3 per m^2.

   (ii) (a)   Explain what is meant by a Type II error in this context.

        (b)   Find the probability of making a Type II error using the test in part (i).

Summary

Steps in a hypothesis test

- Define the variable
- State H_0 and H_1, then give the distribution of the test statistic assuming H_0 is true
- State the rejection rule – you could give a probability statement or define the critical region
- Find whether the test value lies in the critical region – this will usually involve finding a probability and comparing it with the significance level
- Make your conclusion in statistical terms by saying whether H_0 is rejected or not and then **relate it to the situation**

Critical region

If the test value is in the critical region, reject H_0.

If the test value is not in the critical region, do not reject H_0.

Test for binomial proportion p (at the 5% significance level)

	Hypotheses	Critical (rejection) region	Rejection rule (x is the test value)
To test for an **increase** in p	$H_0: p = p_0$ $H_1: p > p_0$	Upper tail 5%	Reject H_0 if $P(X \geqslant x) < 5\%$
To test for a **decrease** in p	$H_0: p = p_0$ $H_1: p < p_0$	Lower tail 5%	Reject H_0 if $P(X \leqslant x) < 5\%$
To test for a **change** in p	$H_0: p = p_0$ $H_1: p \neq p_0$	Upper tail 2.5% Lower tail 2.5%	Reject H_0 if $P(X \geqslant x) < 2.5\%$ or $P(X \leqslant x) < 2.5\%$

Test for Poisson mean λ (at the 5% significance level)

	Hypotheses	Critical (rejection) region	Rejection rule (x is the test value)
To test for an **increase** in λ	$H_0: \lambda = \lambda_0$ $H_1: \lambda > \lambda_0$	Upper tail 5%	Reject H_0 if $P(X \geqslant x) < 5\%$
To test for a **decrease** in λ	$H_0: \lambda = \lambda_0$ $H_1: \lambda < \lambda_0$	Lower tail 5%	Reject H_0 if $P(X \leqslant x) < 5\%$
To test for a **change** in λ	$H_0: \lambda = \lambda_0$ $H_1: \lambda \neq \lambda_0$	Upper tail 2.5% Lower tail 2.5%	Reject H_0 if $P(X \geqslant x) < 2.5\%$ or $P(X \leqslant x) < 2.5\%$

Type I and Type II errors

A Type I error is made when the null hypothesis is rejected but it is in fact true.

A Type II error is made when the null hypothesis is accepted but it is in fact false.

$P(\text{Type I error}) = P(\text{reject } H_0 \text{ when } H_0 \text{ is true})$

$P(\text{Type II error}) = P(\text{accept } H_0 \text{ when } H_1 \text{ is true})$

To find the probability of making a Type II error you must be given a specific value for the alternative hypothesis H_1.

Mixed Exercise 5

1 In a certain country, the postal service director claims that 90% of letters are delivered the day after posting. However, it is suspected that the proportion is less than 90%

When 17 letters are posted at randomly chosen locations, it is found that 13 arrive the next day.

A hypothesis test is carried out at the 5% significance level to test the postal service director's claim.

(i) Write down suitable null and alternative hypotheses for the test.

(ii) Carry out the test, stating your conclusion clearly.

(iii) State the critical region for the test.

2 Over a period of time it is found that the number of defects in a 1 m length of a certain type of steel pipe has mean 2.4

It is suggested that a Poisson distribution may be an appropriate model for the number of defects in a randomly chosen 1 m length of this type of steel pipe.

(i) State two assumptions that would need to be made for the Poisson distribution to be an appropriate model in this case.

A new process is introduced and it is suspected that this may be producing more defects than before. In a quality control experiment, a randomly selected 1 m length of steel pipe is found to have 6 defects.

(ii) Assuming that a Poisson model is appropriate, test, at the 5% significance level, whether the mean number of defects in this type of steel pipe has increased. State your hypotheses clearly.

3 The number of accidents per month at a certain road junction has a Poisson distribution with mean 4.8. A new road sign is introduced warning drivers of the danger ahead, and in a subsequent month 2 accidents occured.

(i) A hypothesis test at the 10% level is used to determine whether there were fewer accidents after the new road sign was introduced. Find the critical region for this test and carry out the test.

(ii) Find the probability of a Type I error.
Cambridge Paper 7 Q4 N02

4 The random variable X is distributed Po(λ). A hypothesis test at the 1% level is to be carried out to test the null hypothesis $\lambda = 0.8$ against the alternative hypothesis $\lambda > 0.8$

The critical region is $X \geqslant c$. Find the value of c.

5 Over many years it is found that, at a particular station, 20% of trains arrive late. A consumer group wishes to test whether the percentage of trains that arrive late has increased recently. It decides to observe 20 trains, randomly selected from the schedule. If more than 4 of the trains arrive late it will claim that the percentage of trains arriving late has increased.

(i) In the case where the percentage of trains arriving late has remained at 20%, find

the probability that the consumer group makes a Type I error.

(ii) In the case where the percentage of trains arriving late has increased to 25%, find the probability that the consumer group makes a Type II error.

6 (i) Explain what is meant by:
    (a) a Type I error,
    (b) a Type II error.

(ii) Roger thinks that a box contains 6 screws and 94 nails. Felix thinks that the box contains 30 screws and 70 nails. In order to test these assumptions they decide to take 5 items at random from the box and inspect them, replacing each item after it has been inspected, and accept Roger's hypothesis (the null hypothesis) if all 5 items are nails.
    (a) Calculate the probability of a Type I error.
    (b) If Felix's hypothesis (the alternative hypothesis) is true, calculate the probability of a Type II error.
Cambridge Paper 7 Q6 N03

7 In a research laboratory where plants are studied, the probability of a certain type of plant surviving was 0.35. The laboratory manager changed the growing conditions and wished to test whether the probability of a plant surviving had increased.

(i) The plants were grown in rows, and when the manager requested a random sample of 8 plants to be taken, the technician took all 8 plants from the front row. Explain what was wrong with the technician's sample.

(ii) A suitable sample of 8 plants was taken and 4 of these 8 plants survived. State whether the manager's test is one-tailed or two-tailed and also state the null and alternative hypotheses. Using a 5% significance level, find the critical region and carry out the test.

(iii) State the meaning of a Type II error in the context of the test in part (ii).

(iv) Find the probability of a Type II error for the test in part (ii) if the probability of a plant surviving is now 0.4
Cambridge Paper 7 Q7 N04

8 At a certain airport 20% of people take longer than an hour to check in. A new computer system is installed, and it is claimed that this will reduce the time to check in. It is decided to accept the claim if, from a random sample of 22 people, the number taking longer than an hour to check in is either 0 or 1.

(i) Calculate the significance level of the test.

(ii) State the probability that a Type I error occurs.

(iii) Calculate the probability that a Type II error occurs if the probability that a person takes longer than an hour to check in is now 0.09.

Cambridge Paper 7 Q4 J07

9 Pieces of metal discovered by people using metal detectors are found randomly in fields in a certain area at an average rate of 0.8 pieces per hectare. People using metal detectors in this area have a theory that ploughing the fields increases the average number of pieces of metal found per hectare. After ploughing, they tested this theory and find that a randomly chosen field of area 3 hectares yielded 5 pieces of metal.

(i) Carry out the test at the 10% level of significance.

(ii) What would your conclusion have been if you had tested at the 5% level of significance?

Jack decides that he will reject the null hypothesis that the average number is 0.8 pieces per hectare if he finds 4 or more pieces of metal in another ploughed field of area 3 hectares.

(iii) If the true mean after ploughing is 1.4 pieces per hectare, calculate the probability that Jack makes a Type II error.

Cambridge Paper 7 Q6 N06

10 Every month Susan enters a particular lottery. The lottery company states that the probability, p, of winning a prize is 0.0017 each month. Susan thinks that the probability of winning is higher than this, and carries out a test based on her 12 lottery results in a one-year period. She accepts the null hypothesis $p = 0.0017$ if she has no wins in the year and accepts the alternative hypothesis $p > 0.0017$ if she wins a prize in at least one of the 12 months.

(i) Find the probability of the test resulting in a Type I error.

(ii) If in fact the probability of winning a prize each month is 0.0024, find the probability of the test resulting in a Type II error.

(iii) Use a suitable approximation, with $p = 0.0024$, to find the probability that in a period of 10 years Susan wins a prize exactly twice.

Cambridge Paper 7 Q5 N08

6 Hypothesis tests 2: z-tests

In this chapter you will learn about

- the language of hypothesis testing related to continuous variables
  - null and alternative hypotheses
  - one-tail and two-tail tests
  - significance level
  - critical (rejection) region
  - test statistic
  - rejection rule
- formulating hypotheses and carrying out a hypothesis test
  - for the population mean where the population is normally distributed with known variance or where a large sample is used
  - for the parameter p of a binomial distribution using a normal approximation (large sample)
  - for the mean λ of a Poisson distribution using a normal approximation (large λ)
- Type I and Type II errors related to these tests

INTRODUCTION TO z-TESTS

In Chapter 5 hypothesis tests relating to discrete distributions were studied.

In this chapter, hypothesis tests are carried out using the **normal distribution**, including the **normal approximation** to the binomial and Poisson distributions. These tests are often referred to as **z-tests**.

The language and methods used in hypothesis testing are illustrated in the following example.

Defining the problem

Ice packs are produced for use in cool boxes. A machine fills the packs with a liquid and is set so that the volume of liquid dispensed follows a normal distribution with mean 524 ml and standard deviation 3 ml. When the ice packs are frozen, the liquid expands. To allow space for this, it is important that the packs are not over-filled.

The machine breaks down and is repaired. When the next batch of ice packs is produced, the supervisor suspects that the mean volume of liquid being dispensed into the packs has increased. This could prove expensive, not only because of the cost of the extra liquid dispensed, but also because the packs might crack when frozen.

To investigate her suspicion, the supervisor decides to take a random sample of 50 packs from the production line. She calculates the mean volume of liquid in a pack for the packs in the sample and finds it to be 524.9 ml.

What does she conclude? Does this provide statistical evidence that the machine is over-filling the packs?

Describing the test

The mean volume of the sample, 524.9 ml, is higher than the required mean of 524 ml. But is it high enough to say that the mean volume of *all* the packs filled by the machine has increased? Perhaps the mean is still 524 ml and this higher value has occurred just because of sampling variation.

To decide, the supervisor needs to carry out a **hypothesis** (or **significance**) test. She will then be able to make a decision that is backed by statistical theory, not just based on a suspicion.

You will notice that the following explanation of carrying out the hypothesis test is similar to that outlined in Chapter 5 (page 135).

Defining the variable

Let X be the volume, in millilitres, of liquid dispensed into a pack after the machine has been repaired.

The mean of X is μ, where μ is unknown. In order to carry out the test you have to assume that the standard deviation remains unchanged.

So $X \sim N(\mu, \sigma^2)$ with $\sigma = 3$

The hypotheses

The **null hypothesis**, denoted by H_0, is that the mean volume dispensed is the same as it was before the repair, i.e. that the mean is 524 ml.

i.e. $H_0: \mu = 524$

Since it is suspected that the mean volume has *increased*, the **alternative hypothesis**, H_1, is that the mean is *greater than* 524 ml.

i.e. $H_1: \mu > 524$ H_1 will contain one of these signs $>$ or $<$ or $\neq$

The test statistic

To carry out the test, the focus moves from X, the volume of liquid in a pack, to the distribution of $\overline{X}$, the **mean volume of a sample of 50 packs**.

$\overline{X}$ is the **test statistic** and you need to know how it is distributed.

In Chapter 4 (page 99) you saw that

if $X \sim N(\mu, \sigma^2)$

then, for samples of size n,

$$\overline{X} \sim N\left(\mu, \frac{\sigma^2}{n}\right)$$

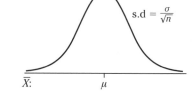

This is the distribution of the sample mean.

The hypothesis test starts by assuming that the value stated in the null hypothesis is true.

If H_0 is true, then $\mu = 524$

Since $\sigma = 3$ and $n = 50$,

$$\overline{X} \sim N\left(524, \frac{3^2}{50}\right)$$

So $\overline{X}$ follows a normal distribution with mean 524 ml and standard deviation $\dfrac{3}{\sqrt{50}}$ ml.

The significance level and type of test

The mean of the random sample of 50 packs is 524.9 ml. This is the **test value**. The outcome of the hypothesis test depends on how close this test value is to 524, the mean stated in the null hypothesis.

If it is *close to* 524 ml, then it could easily have come from a distribution with mean 524 ml and there would not be enough evidence to say that the previous mean of 524 ml has increased.

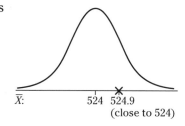

But what is considered close?

If it is *far away from* 524 ml, i.e. in the right-hand (upper) tail of the distribution, then it is unlikely to have come from a distribution with mean 524 ml and it is likely that the mean after repair is higher than 524 ml.

Note that the *upper tail* is being used because the supervisor suspects that there is an *increase* in μ. This type of test is called a **one-tail (upper-tail) test**.

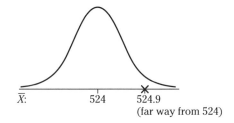

A decision needs to be taken about the cut-off point, c, known as the **critical value**, which indicates the boundary of the region where values of $\bar{X}$ would be considered to be *too far away* from 524 ml and therefore would be *unlikely to occur*. The region is known as the **critical region** or **rejection region**.

The critical value and region are fixed using probabilities linked to the **significance level** of the test. The most commonly used levels of significance are 10%, 5% and 1%.

In general, for an upper-tail test at the α% level, the critical value c is fixed so that $P(\bar{X} > c) = \alpha\%$, and the critical region is $\bar{X} > c$

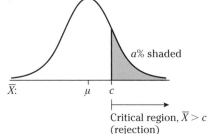

The rejection rule

The hypothesis test involves finding whether or not the sample value, $\bar{x}$, lies in the critical region of the distribution of $\bar{X}$.

- **If $\bar{x}$ lies in the critical region**, a decision is taken that it is too far away from 524 ml to have come from a distribution with this mean. In statistical language, you would **reject the null hypothesis, H_0** (that the mean is 524 ml), **in favour of the alternative hypothesis, H_1** (that the mean is greater than 524 ml).

- **If $\bar{x}$ does not lie in the critical region**, there is not enough evidence to reject H_0, so **H_0 is accepted**.

Values that are not in the critical (rejection) region are in the **acceptance region**, so, in this example, $\bar{X} < c$ is the acceptance region.

For a significance level of α%, if the sample mean lies in the critical (rejection) region, then the result is said to be **significant at the α% level**. Note that if a result is significant at, say, the 1% level, then it is automatically significant at any level greater than 1%, for example 5% or 10%.

Suppose that the supervisor chooses a significance level of 5%.

She will then reject H_0 if the test value (i.e. the mean volume of the sample of 50 cans) lies in the upper-tail 5% of the distribution of sample means.

Since the distribution of $\bar{X}$ is normal, instead of finding c (the critical value of $\bar{x}$), it is possible to work in standardised values and find the z-value that gives 5% in the upper tail.

Using the normal distribution table (page 199)

if $P(Z > z) = 0.05$

then $P(Z < z) = 0.95$

i.e. $\Phi(z) = 0.95$

$z = \Phi^{-1}(0.95)$

$= 1.645$

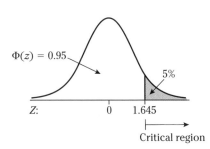

The critical z-value is 1.645

Any z-value greater than 1.645 lies in the upper-tail 5% of the distribution.

This enables a statement to be made, known as the **rejection rule**, which tells you when to reject the null hypothesis.

The **standardised test statistic** is

$$Z = \frac{\overline{X} - \mu}{\frac{\sigma}{\sqrt{n}}}, \text{ where } Z \sim N(0, 1)$$

The rejection rule is:

Reject H_0 if $z > 1.645$, where z is the standardised value of the mean of the sample of 50 packs,

i.e. $z = \dfrac{\overline{x} - \mu}{\frac{\sigma}{\sqrt{n}}} = \dfrac{\overline{x} - 524}{\frac{3}{\sqrt{50}}}$ $\bar{x}$ is the sample mean.

Note: To avoid being influenced by sample readings, the rejection rule is decided *before* the sample is taken.

Does the test value lie in the critical region?

When the sample is taken it is found that the test value, $\overline{x}$, is 524.9

This is standardised and compared with the critical z-value.

So $z = \dfrac{524.9 - 524}{\frac{3}{\sqrt{50}}} = 2.121... > 1.645$

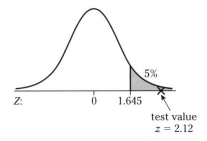

The test value lies in the critical (rejection) region.

Conclusion

The result of the test is now stated in statistical terms and related to the situation as follows.

Since $z > 1.645$, H_0 is rejected in favour of H_1.

The supervisor would conclude that there is evidence that the mean volume of liquid being dispensed by the machine is not 524 ml, but has increased. She would be wise therefore to stop production so that the settings on the machine can be adjusted.

Steps in a hypothesis test for *z*-tests

Here are the steps in a z-test.

- Define the variable
- State H_0 and H_1, then give the distribution of the test statistic assuming H_0 is true
- State the rejection rule – this is usually given in terms of the critical z-value
- Find whether the test value lies in the critical region – this usually involves finding a z-value and comparing it with the critical z-value
- Make your conclusion in statistical terms by saying whether H_0 is rejected or not and then relate it to the situation

The following shows the working you should write down when carrying out this test. The steps are shown in the margin.

Hypothesis test to test whether the machine is over-filling the ice packs

Define the variable
X is the volume, in ml, of liquid in an ice pack. Then $X \sim N(\mu, 3^2)$

State H_0 and H_1 and the distribution of $\overline{X}$
H_0: $\mu = 524$ (The volume is unchanged)

H_1: $\mu > 524$ (The volume has increased)

If H_0 is true, then for samples of size 50,

$$\overline{X} \sim N\left(524, \frac{3^2}{50}\right) \qquad \mu = 524, \frac{\sigma}{\sqrt{n}} = \frac{3}{\sqrt{50}}$$

State the rejection rule

Using a one-tail (upper tail) test at the 5% significance level, the critical z-value is 1.645.

So reject H_0 if $z > 1.645$, where $z = \dfrac{\bar{x} - \mu}{\frac{\sigma}{\sqrt{n}}}$.

Calculate a z-value and compare it with the critical value

The sample mean, $\bar{x} = 524.9$,

so $\quad z = \dfrac{524.9 - 524}{\frac{3}{\sqrt{50}}}$

$\quad\quad = 2.121... > 1.645$

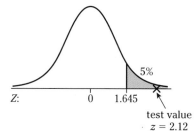

5%

Z: 0   1.645

test value
$z = 2.12$

Remember to show the comparison between the test statistic and the critical value.

Make your conclusion, relating it to the situation.

Since $z > 1.645$, reject H_0 in favour of H_1.

There is evidence, at the 5% level, that the mean volume of liquid being dispensed into the ice packs has increased, so the supervisor's suspicion is confirmed.

Finding the critical value for $\bar{x}$

The critical value c can be found by de-standardising the critical z-value, as follows:

$\dfrac{c - 524}{\frac{3}{\sqrt{50}}} = 1.645$

$c - 524 = 1.645 \times \dfrac{3}{\sqrt{50}}$

$c = 524 + 1.645 \times \dfrac{3}{\sqrt{50}}$

$\quad = 524.69...$

$\quad = 524.7 \ (1 \text{ d.p.})$

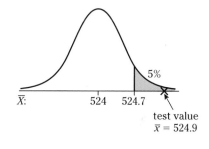

5%

$\bar{X}$:   524   524.7

test value
$\bar{x} = 524.9$

The critical value is 524.7 ml and the critical region is $\bar{X} > 524.7$ ml. This means that any test value greater than 524.7 ml lies in the critical region.

Since the supervisor's test value of 524.9 ml is greater than 524.7 ml, it lies in the critical region, confirming the result obtained above.

One-tail and two-tail tests

Suppose that the null hypothesis is

$\quad$ H_0: $\mu = \mu_0$ $\quad\quad\quad$ μ_0 is a specified value.

One-tail test

You will recall from chapter 5 (page 140) that in a **one-tail** test, the alternative hypothesis H_1 looks for an **increase** or a **decrease** in μ.

If H_1 looks for an **increase**, then H_1 is $\mu > \mu_0$ and the critical region is in the **upper tail**.

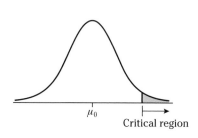

μ_0

Critical region

If H_1 looks for a **decrease**, then H_1 is $\mu < \mu_0$ and the critical region is in the **lower tail**.

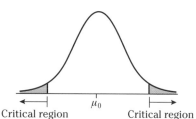

Critical region μ_0

Two-tail test

In a **two-tail test**, the alternative hypothesis H_1 looks for a **change** in μ without specifying whether it is an increase or a decrease.

The alternative hypothesis H_1 is $\mu \neq \mu_0$.

The critical region is in two parts, symmetrically placed in the lower and upper tails.

Critical region $\quad \mu_0 \quad$ Critical region

Critical *z*-values

Critical values depend on the significance level of the test and whether the test is one-tail or two-tail. The most commonly used values are given in the table below, which is provided in the examination. The table is also printed on page 199.

Critical values for the normal distribution

If Z has a normal distribution with mean 0 and variance 1 then, for each value of p, the table gives the value of z such that $P(Z < z) = p$

p	0.75	0.90	0.95	0.975	0.99	0.995	0.9975	0.999	0.9995
z	0.674	1.282	1.645	1.960	2.326	2.576	2.807	3.090	3.291

One-tail test

For example, for a **one-tail test** at the **1% level**, you want to find z such that $\Phi(z) = 0.99$, so look up $p = 0.99$

From the table, $z = 2.326$

For an upper-tail test, the critical z-value is 2.326 and the critical region is $z > 2.326$

For a lower-tail test, by symmetry the critical z-value is -2.326 and the critical region is $z < -2.326$

Upper tail   **Lower tail**

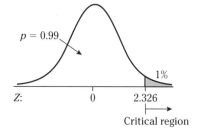

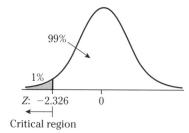

Two-tail test

In a **two-tail test** at the **1% level**, the area of 1% is split evenly between the upper and lower tails, with 0.5% in each, i.e. 0.005. There are two critical values.

To find the upper-tail value, you need to find z such that $\Phi(z) = 0.995$, so you look up $p = 0.995$

From the table, $z = 2.576$

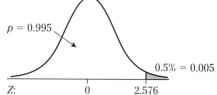

So the upper-tail critical z-value is 2.576

By symmetry, the lower-tail critical z-value is -2.576

The critical region is $z > 2.576$ and $z < -2.576$

This can be written in one statement as $|z| > 2.576$

Note: If the value for the significance level you require is not in the table (for example, 97%), use the main table (headed **The normal distribution function**) in reverse. This is printed on page 199.

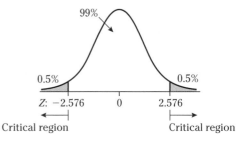

Summary of critical values and rejection rules

The summary below shows the critical z-values and rejection rules for the most commonly used levels of significance of 10%, 5% and 1%.

Significance level	One-tail (lower-tail) $H_0: \mu = \mu_0$ $H_1: \mu < \mu_0$	One-tail (upper-tail) $H_0: \mu = \mu_0$ $H_1: \mu > \mu_0$	Two-tail $H_0: \mu = \mu_0$ $H_1: \mu \neq \mu_0$
10%	Reject H_0 if $z < -1.282$	Reject H_0 if $z > 1.282$	Reject H_0 if $z > 1.645$ or $z < -1.645$ i.e. if $|z| > 1.645$
5%	Reject H_0 if $z < -1.645$	Reject H_0 if $z > 1.645$	Reject H_0 if $z > 1.96$ or $z < -1.96$ i.e. if $|z| > 1.96$
1%	Reject H_0 if $z < -2.326$	Reject H_0 if $z > 2.326$	Reject H_0 if $z > 2.576$ or $z < -2.576$ i.e. if $|z| > 2.576$

HYPOTHESIS TEST 1: TESTING μ, THE MEAN OF A POPULATION

Consider a population X with mean μ (unknown) and variance σ^2. A value for μ, call it μ_0, is specified in the null hypothesis, for example:

$H_0: \mu = \mu_0$

$H_1: \mu < \mu_0$ (or $\mu > \mu_0$ or $\mu \neq \mu_0$)

To test the hypotheses, take a random sample of size n from the population and calculate the sample mean. The test statistic is $\overline{X}$ and you use the distribution of the sample mean.

As with confidence intervals (studied in Chapter 4), there are now several cases that could occur, depending on whether or not the population is normal, whether the sample size is large or small and whether or not the population variance is known.

Case 1

We first examine the case of a hypothesis test for population mean μ when

- the population variable X is **normally distributed**
- the population variance σ^2 is **known**
- the sample can be **any size**, large or small.

Since the population is normally distributed, $X \sim N(\mu, \sigma^2)$

The distribution of sample means is also normally distributed *for all sample sizes*, with mean μ_0, the value specified in the null hypotheses H_0.

For a hypothesis test for the mean μ of a normal population X with known variance σ^2 based on a sample of any size:

The test statistic is $\overline{X}$, where

$$\overline{X} \sim N\left(\mu, \frac{\sigma^2}{n}\right)$$

In standardised form, the test statistic is

$$Z = \frac{\overline{X} - \mu}{\frac{\sigma}{\sqrt{n}}} \qquad \text{where } Z \sim N(0, 1)$$

The value of μ is specified in the null hypothesis.

Example 6.1

Each year a large number of students taking a particular course at a college sit an examination. Over a period of time it is found that the marks of the students at this college follow a normal distribution with mean 70 and standard deviation 6.

This year the examination contains questions on a new topic and a lecturer believes that the marks are lower, on average. To test this belief, he calculates the mean mark of a random sample of 25 students and finds it to be 67.3

Carry out a hypothesis test at the 5% level of significance to test whether the mean mark in the examination is lower this year. Assume that the marks are normally distributed and the standard deviation is 6.

X is the examination mark of a student, where $X \sim N(\mu, 6^2)$.

$H_0: \mu = 70$ (The mean mark is unchanged)

$H_1: \mu < 70$ (The mean mark is lower this year)

If H_0 is true then, for samples of size 25,

$$\overline{X} \sim N\left(70, \frac{6^2}{25}\right) \qquad \mu = 70, \quad \frac{\sigma}{\sqrt{n}} = \frac{6}{\sqrt{25}}$$

You are looking for a decrease in μ.

Use a one-tail (lower tail) test at the 5% significance level.

Find the critical z-value that gives 5% in the lower tail.

The critical z-value is negative, where

$$\Phi(-z) = 0.95,$$
$$-z = 1.645$$
$$z = -1.645$$

Reject H_0 if $z < -1.645$, where $z = \dfrac{\overline{x} - \mu}{\frac{\sigma}{\sqrt{n}}}$.

$$\overline{x} = 67.3,$$

so
$$z = \frac{67.3 - 70}{\frac{6}{\sqrt{25}}}$$

$$= -2.25 < -1.645$$

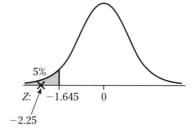

Put a cross on the sketch to show z.

Since $z < -1.645$, reject H_0 in favour of H_1.

There is evidence, at the 5% level, that the mean examination mark is lower this year, supporting the lecturer's belief.

Calculation note

You could calculate the value of $\frac{\sigma}{\sqrt{n}}$, where $\frac{\sigma}{\sqrt{n}} = \frac{6}{\sqrt{25}} = 1.2$, but it is often useful to leave it in its

uncalculated form as a reminder that the distribution of the **sample mean** is being used.

Finding the critical region

To find the critical region, calculate the critical value of $\bar{x}$ as follows:

$$\frac{c - 70}{\frac{6}{\sqrt{25}}} = -1.645$$

$$c - 70 = -1.645 \times \frac{6}{\sqrt{25}}$$

$$c = 70 - 1.645 \times \frac{6}{\sqrt{25}}$$

$$c = 68.026$$

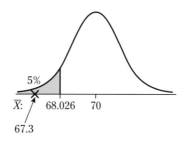

The critical value is 68.026 and the critical region is $\bar{X} < 68.026$

This means that any test value less than 68.026 would result in the null hypothesis being rejected.

Example 6.2

In Europe the diameters of women's rings have mean 18.5 mm. Researchers claim that women in Jakarta have smaller fingers than women in Europe. The researchers took a random sample of 20 women in Jakarta and measured the diameters of their rings. The mean diameter was found to be 18.1 mm. Assuming that the diameters of women's rings in Jakarta have a normal distribution with standard deviation 1.1 mm, carry out a hypothesis test at the $2\frac{1}{2}\%$ level to determine whether the researchers' claim is justified.

Cambridge Paper 7 Q1 J09(71)

X is the diameter, in mm, of a woman's ring in Jakarta, where $X \sim N(\mu, 1.1^2)$

H_0: $\mu = 18.5$ (The mean diameter is the same as in Europe)

H_1: $\mu < 18.5$ (The mean diameter is less than in Europe)

If H_0 is true then, for samples of size 20,

$$\bar{X} \sim N\left(18.5, \frac{1.1^2}{20}\right) \qquad \mu = 18.5, \frac{\sigma}{\sqrt{n}} = \frac{1.1}{\sqrt{20}}$$

Use a one-tail (lower tail) test at the $2\frac{1}{2}\%$ significance level.

The critical z-value is negative, where

$$\Phi(-z) = 0.975,$$

so $-z = 1.96$

$$z = -1.96$$

Reject H_0 if $z < -1.96$, where $z = \frac{\bar{x} - \mu}{\frac{\sigma}{\sqrt{n}}}$.

$$\bar{x} = 18.1,$$

so $$z = \frac{18.1 - 18.5}{\frac{1.1}{\sqrt{20}}}$$

$$= -1.626...... > -1.96$$

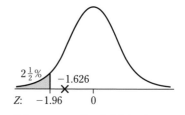

Put a cross on the sketch to show the comparison.

Since $z > -1.96$, do not reject H_0.

There is not enough evidence, at the $2\frac{1}{2}\%$ level, that the mean diameter of women's rings in Jakarta is less than in Europe. So there is not enough evidence to support the researchers' claim that the women in Jakarta have smaller fingers.

Example 6.3

A machine packs flour into bags. A random sample of 11 bags was taken and the masses, in grams, of the bags were:

1506.8, 1506.4, 1506.3, 1507.2, 1506.1, 1506.8, 1506.6, 1507.0, 1507.5, 1506.3, 1506.4

Filled bags are supposed to have a mass of 1506.5 g. Test whether the sample provides evidence, at the 8% significance level, that the machine is producing overweight bags. Assume that the mass of bags of flour has a normal distribution with standard deviation 0.4 g.

X is the mass, in grams, of a bag of flour, where $X \sim N(\mu, 0.4^2)$.

$\quad$ H_0: $\mu = 1506.5$ $\quad$ (The mean mass is 1506.5 g)

$\quad$ H_1: $\mu > 1506.5$ $\quad$ (The mean mass is greater than 1506.5 g)

If H_0 is true then, for samples of size 11,

$$\overline{X} \sim N\left(1506.5, \frac{0.4^2}{11}\right) \qquad \mu = 1506.5, \frac{\sigma}{\sqrt{n}} = \frac{0.4}{\sqrt{11}}$$

Use a one-tail test at the 8% significance level.

Find the critical z-value with 8% in the upper tail.

As $\Phi^{-1}(0.92)$ is not in the critical values table, you need to use the main Normal distribution table in reverse.

$$z = \Phi^{-1}(0.92)$$

$$= 1.419$$

Reject H_0 if $z > 1.419$, where $z = \dfrac{\overline{x} - \mu}{\frac{\sigma}{\sqrt{n}}}$.

Calculate the sample mean by using the calculator in statistical mode:

$\quad$ $\overline{x} = 1506.67...$

So $\qquad z = \dfrac{1506.67... - 1506.5}{\dfrac{0.4}{\sqrt{11}}}$

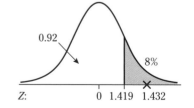

$\qquad\quad = 1.432... > 1.419$

Since $z > 1.419$, reject H_0 in favour of H_1.

There is evidence, at the 8% level, that the mean mass has increased and the machine is producing overweight bags.

Example 6.4

The times taken for pupils in Ming's year group to do their English homework have a normal distribution with standard deviation 15.7 minutes. A teacher estimates that the mean time is 42 minutes. The times taken by a random sample of 3 students from the year group were 27, 35 and 43 minutes. Carry out a hypothesis test at the 10% significance level to determine whether the teacher's estimate for the mean should be accepted, stating the null and alternative hypotheses.

Cambridge Paper 7 Q2 N08

X is the time, in minutes, taken by a student to do the homework, where $X \sim N(\mu, 15.7^2)$.

H_0: $\mu = 42$ (The mean time is 42 minutes)

H_1: $\mu \neq 42$ (The mean time is not 42 minutes)

If H_0 is true then, for samples of size 3,

$$\overline{X} \sim N\left(42, \frac{15.7^2}{3}\right) \qquad \mu = 42, \frac{\sigma}{\sqrt{n}} = \frac{15.7}{\sqrt{3}}$$

This is a two-tail test as you are looking for a change in μ, not specifically an increase or a decrease.

Use a two-tail test at the 10% significance level.

The 10% of the distribution in the critical region is split equally between the upper and lower tail, with 5% in each.

To find the critical z-values, first the upper tail value.

$$z = \Phi^{-1}(0.95)$$

$$= 1.645$$

By symmetry, the lower tail value is -1.645, so the critical *z*-values are ± 1.645.

Reject H_0 if $z > 1.645$ or $z < -1.645$, i.e. if $|z| > 1.645$,

where $z = \dfrac{\overline{x} - \mu}{\frac{\sigma}{\sqrt{n}}}.$

Calculate the sample mean.

$$\overline{x} = \frac{27 + 35 + 43}{3} = 35$$

So $z = \dfrac{35 - 42}{\frac{15.7}{\sqrt{3}}}$

$$= -0.7722\ldots$$

and $|z| = 0.7722\ldots < 1.645$

Note: It is easier to do the comparison if you use $|z|$ rather than z .

Since $|z| < 1.645$, do not reject H_0.

There is not enough evidence, at the 10% level, to reject the teacher's estimate of 42 minutes for the mean time, so this estimate is accepted.

Note: You are not saying that the mean time definitely is 42 minutes, but that you do not have enough evidence to say that it is not 42 minutes.

Example 6.5

A sample of size 16 is taken from the distribution of X, where $X \sim N(\mu, 3^2)$ and a hypothesis test carried out at the 0.5% level of significance. The sample mean is m.

Find the set of values of m which result in the rejection of the null hypothesis $\mu = 100$ in favour of the alternative hypothesis $\mu > 100$

$X \sim N(\mu, 3^2)$

The hypotheses are

$$H_0: \mu = 100 \qquad H_1: \mu > 100$$

If H_0 is true, then, for samples of size 16,

$$\overline{X} \sim N\left(100, \frac{3^2}{16}\right) \qquad \mu = 100, \frac{\sigma}{\sqrt{n}} = \frac{3}{\sqrt{16}}$$

The test is a one-tail (upper-tail) test at the 0.5% level.

As H_0 is rejected, the sample mean m lies in the critical region. To define the critical region, find the critical value c.

First find the corresponding critical z-value.

The critical z-value that gives 0.5% in the upper tail is 2.576 (see page 166).

Now de-standardise the critical z-value to find c.

$$\frac{c - 100}{\dfrac{3}{\sqrt{16}}} = 2.576$$

$$c - 100 = 2.576 \times \frac{3}{\sqrt{16}}$$

$$c = 100 + 2.576 \times \frac{3}{\sqrt{16}}$$

$$= 101.932$$

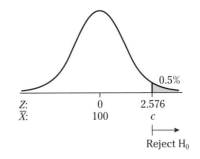

The critical value is 101.932 and the critical region is $\overline{X} > 101.932$

For H_0 to be rejected, m must lie in the critical region, so $m > 101.932$

Exercise 6a

1 In each of the following, X follows a normal distribution with unknown mean μ and known variance σ^2. A random sample of size n is taken from the population of X and the sample mean, $\overline{x}$, is calculated.

Test the hypotheses stated, at the significance level indicated.

	Distribution of X	n	$\overline{x}$	Hypotheses		Significance level
(i)	$X \sim N(\mu, 1.44)$	10	27	$H_0: \mu = 26.3$	$H_1: \mu > 26.3$	5%
(ii)	$X \sim N(\mu, 17.64)$	49	125	$H_0: \mu = 123.5$	$H_1: \mu > 123.5$	1%
(iii)	$X \sim N(\mu, 0.18^2)$	100	4.35	$H_0: \mu = 4.40$	$H_1: \mu < 4.40$	2%
(iv)	$X \sim N(\mu, 3^2)$	30	15.2	$H_0: \mu = 15.8$	$H_1: \mu \neq 15.8$	5%

2 A machine fills cans with soft drinks so that their contents have a nominal volume of 330 ml. Over a period of time it is found that the volume of liquid in the cans follows a normal distribution with mean 335 ml and standard deviation 3 ml.

A setting on the machine is altered, following which the operator suspects that the mean volume of liquid discharged by the machine into the cans has decreased. He takes a random sample of 50 cans and finds that the mean volume of liquid in these cans is 334.6 ml.

Carry out a hypothesis test at the 5% significance level to test whether the mean volume of liquid in the cans has decreased. Assume that the standard deviation is still 3 ml.

3 When clients visit a certain firm of solicitors, the duration of appointments, in minutes, may be modelled by a normal distribution with mean 50 minutes and standard deviation 5.3 minutes.

Following the appointment of two new solicitors, the mean duration of a random sample of 15 appointments was 54.2 minutes. Assuming that the standard deviation is still 5.3 minutes, test, at the 1% significance level, whether the mean duration of appointments has increased.

4 Czarina carries out a test, at the 5% significance level, using a normal distribution. The null hypothesis is $\mu = 103.5$ and the alternative hypothesis is $\mu < 103.5$

Czarina finds that the value of the test statistic is -1.350

What conclusion should Czarina draw?

5 A marmalade manufacturer produces thousands of jars of marmalade each week. The mass of marmalade in a jar may be modelled by a normal distribution with mean 455 g and standard deviation 0.8 g.

Following a slight adjustment to the filling machine, a random sample of 10 jars is found to contain the following masses, in grams, of marmalade.

454.8   453.8   455.0   454.4   455.4
454.4   454.4   455.0   455.0   453.6

Assuming that the standard deviation is unaltered by the adjustment, test at the 5% significance level whether there has been a change in the mean mass of marmalade in jars produced by the manufacturer.

6 The masses of components produced at a particular workshop are normally distributed with standard deviation 0.8 g. It is claimed that the mean mass is 6.0 g.

To test this claim, the mean mass of a random sample of 50 components is calculated and a hypothesis test at the 5% level carried out. On the basis of the test, the claim is accepted.

(i) State suitable null and alternative hypotheses.

(ii) Between what values does the mean mass of the 50 components in the sample lie?

7 An athlete finds that her times for running a race are normally distributed with mean 10.75 s and standard deviation 0.05 s. She trains intensively for a week and then records her time in the next 6 races. Her times, in seconds, are:

10.70, 10.63, 10.75, 10.81, 10.66, 10.72

Is there evidence, at the 5% level, that training intensively has improved her times?

8 A resident of an urban road in the UK claims that the average speed of vehicles using the road is greater than the 30 mph speed limit. To investigate this claim, a random sample of 25 vehicles is taken and the time each takes to travel along a measured mile of the road is measured. It is assumed that the speeds, in miles per hour, calculated from these observations may be modelled by a normal distribution with mean μ and standard deviation 12.

A hypothesis test is carried out to test the null hypothesis H_0: $\mu = 30$ against the alternative hypothesis H_1: $\mu > 30$.

(i) The critical region for a hypothesis test, at the 5% level of significance is $\overline{X} > k$. Find the value of k.

(ii) State, with a reason, your conclusion for the test when the mean speed calculated from the sample was 35 mph.

Case 2

We now examine the case of a hypothesis test for population mean μ when

- the distribution of X is **not normal**
- the population variance σ^2 is **known**
- the sample size is large.

Since the population of X is not normal, you cannot say that the distribution of the test statistic $\overline{X}$ is normal for all sample sizes.

However, if the sample size n is **large** ($n \geqslant 30$, say), you can apply the Central Limit theorem (see page 105).

By the **Central Limit theorem**, if samples of size n taken from any non-normal population X, the distribution of $\overline{X}$ is **approximately normal**, provided that n is large ($n \geqslant 30$, say).

For a hypothesis test for the mean μ of a non-normal population X with known variance σ^2, based on the mean of a large sample ($n \geqslant 30$, say):

The test statistic is $\overline{X}$, where

$$\overline{X} \sim N\left(\mu, \frac{\sigma^2}{n}\right) \text{ approximately.}$$

> The value of μ is specified in the null hypothesis.

In standardised form, the test statistic is

$$Z = \frac{\overline{X} - \mu}{\frac{\sigma}{\sqrt{n}}} \text{ where } Z \sim N(0, 1)$$

Example 6.6

The manager of a large hospital states that the mean age of patients at the hospital is 45 years. Records of a random sample of 100 patients at the hospital give a mean age of 48.4 years. Assuming that the population standard deviation is 18.0 years, test, at the 5% level of significance, whether the manager's statement should be accepted. State clearly the null and alternative hypotheses.

X is the age, in years, of a patient. The distribution of X is unknown. The mean of X is μ and the standard deviation is σ.

H_0: $\mu = 45$ (The mean is 45)

H_1: $\mu \neq 45$ (The mean is not 45)

The sample size is large ($n = 100$), so the Central Limit theorem can be used.

By the Central Limit theorem,

if H_0 is true then, for samples of size 100,

$$\overline{X} \sim N\left(45, \frac{18.0^2}{100}\right) \text{ approx.} \qquad \mu = 45, \frac{\sigma}{\sqrt{n}} = \frac{18.0}{\sqrt{100}}$$

Use a two-tail test at the 5% level.

> This is a two-tail test as you are not looking specifically for an increase or a decrease in μ.

The critical z-values for a two-tail test at the 5% level are ± 1.96 (see page 167).

Reject H_0 if $z < -1.96$ or $z > 1.96$, i.e. if $|z| > 1.96$,

where $$z = \frac{\bar{x} - \mu}{\frac{\sigma}{\sqrt{n}}}.$$

You are given that $\bar{x} = 48.4$,

so $$z = \frac{48.4 - 45}{\frac{18.0}{\sqrt{100}}}$$

$$= 1.888$$

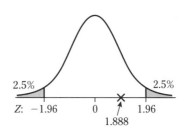

and $$|z| = 1.888 < 1.96.$$

Since $|z| < 1.96$, do not reject H_0.

There is not enough evidence, at the 5% level, to say that the mean age is not 45 years, so the manager's statement that the mean age is 45 years is accepted.

Case 3

Finally, we examine the case of a hypothesis test for population mean μ when

- the population variance σ^2 is **unknown**
- the sample size is **large**.

When carrying out hypothesis test it is often the case that the population variance σ^2 is unknown. Provided that the sample size is large ($n \geqslant 30$, say), it is permissible to use s^2 instead, where s^2 is the **unbiased estimate** of σ^2.

You will recall from Chapter 4 (page 109) that

$$s^2 = \frac{1}{n-1}\left(\Sigma x^2 - \frac{(\Sigma x)^2}{n}\right)$$ The formula for s^2 is given in the examination.

For a hypothesis test for the mean μ of population X with **unknown variance** σ^2, based on the mean of a large sample, size $n \geqslant 30$:

The test statistic is $\bar{X}$, where

$$\bar{X} \sim N\left(\mu, \frac{s^2}{n}\right) \text{ approximately,} \quad \text{where } s^2 = \frac{1}{n-1}\left(\Sigma x^2 - \frac{(\Sigma x)^2}{n}\right)$$

In standardised form, the test statistic is

$$Z = \frac{\bar{X} - \mu}{\frac{s}{\sqrt{n}}} \quad \text{where } Z \sim N(0, 1)$$ The value of μ is specified in the null hypothesis.

Example 6.7

The manufacturer of a certain type of light bulb claims that the average lifetime of the firm's light bulbs is 1000 hours. A consumer group thinks that the manufacturer is overstating the lifetime of the light bulbs and tests a random sample of 64 bulbs, recording the life x, in hours, of each bulb.

The results are summarised below:

$\Sigma x = 63\,910.4$ and $\Sigma x^2 = 63\,824\,061$

(i) Calculate the sample mean.

(ii) Calculate an unbiased estimate of the standard deviation of the lifetime of this type of light bulb.

(iii) Is there evidence, at the 10% level, that the manufacturer is overstating the lifetime of light bulbs produced by the firm?

(iv) State whether it was necessary to use the Central Limit theorem in your answer in part (iii) and if it was, state at what point it was necessary to use it.

(i) $\bar{x} = \dfrac{\Sigma x}{n}$

$= \dfrac{63910.4}{64}$

$= 998.6$

(ii) $s^2 = \dfrac{1}{n-1}\left(\Sigma x^2 - \dfrac{(\Sigma x)^2}{n}\right)$

$= \dfrac{1}{63}\left(63824061 - \dfrac{63910.4^2}{64}\right)$

$= 49.77\ldots$

$s = \sqrt{49.77}$

$= 7.0548\ldots$

$= 7.04 \ (3 \text{ s.f.})$

(iii) X is the lifetime, in hours, of a light bulb. The mean is μ and the standard deviation is σ. The distribution of X is unknown.

$H_0: \mu = 1000$ (The mean is 1000 h)

$H_1: \mu < 1000$ (The mean is less than 1000 h and the manufacturer is overstating the lifetime)

Since n is large ($\geqslant 30$), by the Central Limit theorem

$\bar{X} \sim N\left(\mu, \dfrac{s^2}{n}\right)$ approximately   Since σ^2 is unknown, s^2 is used instead.

If H_0 is true then

$\bar{X} \sim N\left(1000, \dfrac{49.77\ldots}{64}\right)$ $\mu = 1000, \dfrac{s}{\sqrt{n}} = \dfrac{7.054\ldots}{\sqrt{64}}$

The test is a one-tail (lower tail) test at the 10% level.

The critical z-value is negative, where

$\Phi(-z) = 0.9,$

so $-z = 1.282$

$z = -1.282$

Reject H_0 if $z < -1.282$, where $z = \dfrac{\bar{x} - \mu}{\frac{s}{\sqrt{n}}}$

$\bar{x} = 998.6,$

so $z = \dfrac{998.6 - 1000}{\frac{7.054\ldots}{\sqrt{64}}}$

$= -1.587 < -1.282$

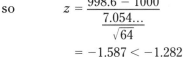

Since $z < -1.282$, reject H_0 in favour of H_1.

There is evidence, at the 10% level, that the mean is less than 1000 hours and the manufacturer is overstating the length of lifetime of the light bulbs produced by the firm.

(iv) It was necessary to use the Central Limit theorem because you do not know whether the lifetimes of the bulbs are normally distributed. The Central Limit theorem tells you that, in this situation, the distribution of the mean lifetimes of the bulbs is approximately normal. This is needed to specify the distribution of the test statistic $\overline{X}$, as stated in part (iii).

Example 6.8

From previous years' observations, the lengths of salmon in a river were found to be normally distributed with mean 65 cm. A researcher suspects that pollution in water is restricting growth. To test this theory, she measures the length x cm of a random sample of n salmon and calculates that $\overline{x} = 64.3$ and $s = 4.9$, where s^2 is the unbiased estimate of the population variance. She then carries out an appropriate hypothesis test.

(i) Her test statistic z has a value of -1.807 correct to 3 decimal places. Calculate the value of n.

(ii) Using this test statistic, carry out the hypothesis test at the 5% level of significance and state what her conclusion should be.

Cambridge Paper 7 Q3 N02

(i) X is the length, in centimetres, of a salmon, where $X \sim N(\mu, \sigma^2)$

Since σ^2 is unknown, s^2 is used instead, where $s = 4.9$

H_0: $\mu = 65$ (The mean is unchanged)

H_1: $\mu < 65$ (The mean has decreased)

Since n is large, by the Central Limit theorem,

$$\overline{X} \sim N\left(\mu, \frac{s^2}{n}\right) \text{ approximately.}$$

If H_0 is true, then

$$\overline{X} \sim N\left(65, \frac{4.9^2}{n}\right) \qquad \mu = 65, \frac{s}{\sqrt{n}} = \frac{4.9}{n}, \text{ where } n \text{ is unknown.}$$

The standardised test statistic is

$$Z = \frac{\overline{X} - \mu}{\frac{s}{\sqrt{n}}} = \frac{\overline{X} - 65}{\frac{4.9}{\sqrt{n}}}$$

You are given that the standardised value of $\overline{x} = 64.3$ is $z = -1.807$

so $$\frac{64.3 - 65}{\frac{4.9}{\sqrt{n}}} = -1.807$$

$$64.3 - 65 = -1.807 \times \frac{4.9}{\sqrt{n}}$$

$$-0.7 = -1.807 \times \frac{4.9}{\sqrt{n}}$$

$$\sqrt{n} = \frac{1.807 \times 4.9}{0.7} = 12.649$$

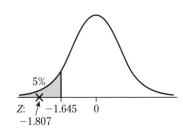

Square both sides:

$$n = (12.649)^2 = 159.997\ldots$$

Since n must be an integer, $n = 160$

(ii) Using a one-tail test at the 5% level, reject H_0 if $z < -1.645$

From part (i), $z = -1.807 < -1.645$, so H_0 is rejected in favour of H_1.

There is evidence, at the 5% level, that the mean length of salmon has decreased, so the researcher would conclude that there is evidence of restricted growth.

Note, however, that this does not prove that the restricted growth is *caused* by pollution in the water. The researcher would need to carry out scientific tests to establish a causal link.

Exercise 6b

1 A random sample of 75 11-year-olds performed a simple task and the time taken, t minutes, noted for each. The results were summarised as follows:

$$\Sigma t = 1215, \quad \Sigma t^2 = 21\,709$$

(i) Calculate an unbiased estimate of the population variance of the time taken to perform the task.

(ii) Test, at the 1% significance level, whether there is evidence that the mean time taken to perform the task is greater than 15 min.

2 An inspector checks items from a production line. On average, she takes 22.5 s to check each item. After the installation of a new lighting system the times, t seconds, that she takes to check each of 50 randomly chosen items from the production line are summarised by

$$\Sigma t = 1107, \quad \Sigma t^2 = 24\,592.35$$

(i) Calculate an unbiased estimate of the population variance of the time taken to check an item under the new lighting system.

(ii) Test at the 2% significance level whether there is evidence that the mean time to inspect an item has been reduced.

(iii) A technician who carried out the above test concluded with the following incorrect statement
'It is not necessary for the population to be normal since the sample size is large and the Central Limit theorem states that any sufficiently large sample is normal.'
Give a corrected version.

3 A continuous variable X has mean μ and variance 32. It is thought that $\mu = 55.0$

The mean of a random sample of 81 observations of the variable is 56.2

Does this provide evidence at the 10% level of significance that the mean is not 55.0?

4 A sample of 40 observations from a normal distribution X gave

$$\Sigma x = 24 \text{ and } \Sigma x^2 = 596$$

Perform a two-tail test, at the 5% significance level, to test whether the mean of the distribution is zero.

5 An intelligence test is developed in which the mean score is 100 and the standard deviation is 12. When the test was given to a random sample of 50 children from a particular city, the mean score was 105. Does this provide evidence, at the 5% level, that children in this city are generally more intelligent?

6 Cans of lemonade are filled by a machine which is set to dispense a mean amount of 330 ml into each can. The manufacturer suspects that the machine is tending to over-dispense and, in order to test the suspicion, measures the contents, x millilitres, of a random sample of 30 cans. The results are summarised by

$$\Sigma x = 9925 \text{ and } \Sigma x^2 = 3\,284\,137$$

(i) Calculate an unbiased estimate of the population variance of the amount dispensed into each can.

(ii) Test the manufacturer's suspicion at the 10% significance level.

(iii) Indicate where the Central Limit theorem is used in the test, and state why the use of the Central Limit theorem is necessary.

7   An ambulance station serves an area which includes more than 10 000 houses. It has been decided that if the mean distance of the houses from the ambulance station is greater than 10 km then a new ambulance station will be necessary. The distance, x kilometres, from the station of each of a random sample of 200 houses was measured, the results being summarised by

$$\Sigma x = 2092.0 \text{ and } \Sigma x^2 = 24\,994.5$$

(i) Calculate, correct to 4 significant figures, unbiased estimates of the population mean μ and the population variance of the distances, in km, of the houses from the station.

A hypothesis test of the null hypothesis $\mu = 10$ against the alternative hypothesis $\mu > 10$ is carried out at the $\alpha\%$ significance level. The critical region for the test is $\overline{X} > 10.65$

(ii) Calculate the value of α.

(iii) State the conclusion of the test.

(iv) Give a reason why it is not necessary for the distances to be normally distributed for the test to be valid.

TYPE I AND TYPE II ERRORS

You will recall from Chapter 5 (page 148) that, when you carry out a hypothesis test, there are four possible conclusions, two of which lead to a correct decision being made and the other two to a wrong decision being made. The errors associated with making wrong decisions are called Type I and Type II errors.

The outcomes and errors are summarised as follows:

- H_0 is true and your test leads you to accept H_0: correct decision
- H_0 is true but your test leads you to reject H_0: wrong decision, Type I error
- H_0 is false but your test leads you to accept H_0: wrong decision, Type II error
- H_0 is false and your test leads you to reject H_0: correct decision.

Type I error

A **Type I error** is made when you reject H_0 but in fact H_0 is true,

i.e.   P(Type I error) = P(reject H_0 when H_0 is true)

When carrying out z-tests, there is a relationship between the probability of making a Type I error and the significance level of the test.

Suppose you are using a one-tail test at the 5% level and H_0 is true.

If the test value falls within the shaded region you would reject H_0, **even though H_0 is true**. You would be making a Type I error.

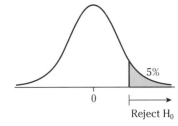

The probability of this Type I error is 5%, which is equal to the level of significance of the test.

So, for a z-test:

        P(Type I error) = significance level of test

Type II error

A **Type II error** is made when you accept H_0 but in fact H_0 is false.

To calculate the probability of making a Type II error, you must be given a specific value for the alternative hypothesis H_1.

Then   P(Type II error) = P(accept H_0 when H_1 is true)

Note: A Type I error involves rejecting H_0. So it follows that if a test leads to H_0 being accepted, then a Type I error cannot be made.

A Type II error involves accepting H_0. So it follows that if a test leads to H_0 being rejected, then a Type II error cannot be made.

Example 6.9

A study of a large sample of books by a particular author shows that the number of words per sentence can be modelled by a normal distribution with mean 21.2 and standard deviation 7.3. A researcher claims to have discovered a previously unknown book by this author. The mean length of 90 sentences chosen at random in this book is found to be 19.4 words.

(i) Assuming the population standard deviation of sentence lengths in this book is also 7.3, test at the 5% level of significance whether the mean sentence length is the same as the author's. State your null and alternative hypotheses.

(ii) State in words relating to the context of the test what is meant by a Type I error and state the probability of a Type I error in the test in part (i).

Cambridge Paper 7 Q4 J05

(i) X is the number of words per sentence in the unknown book, where $X \sim N(\mu, 7.3^2)$

H_0: $\mu = 21.2$ (The mean sentence length is the same as the author's sentence length)

H_1: $\mu \neq 21.2$ (The mean sentence length is not the same as the author's sentence length)

If H_0 is true, then, for samples of size 90,

$$\bar{X} \sim N\left(21.2, \frac{7.3^2}{90}\right) \qquad \mu = 21.2, n = 90, \frac{\sigma}{\sqrt{n}} = \frac{7.3}{\sqrt{90}}$$

Use a two-tail test at the 5% significance level.

The 5% of the distribution in the critical region is split equally between the upper and lower tails, with 2.5% in each.

To find the critical z-values, first find $\Phi^{-1}(0.975)$.

This gives $z = 1.96$, so the critical z-values are ± 1.96

Reject H_0 if $z > 1.96$ or $z < -1.96$ i.e. if $|z| > 1.96$,

where $\qquad z = \dfrac{\bar{x} - 21.2}{\frac{7.3}{\sqrt{90}}}$

The sample mean is 19.4

so $\qquad z = \dfrac{19.4 - 21.2}{\frac{7.3}{\sqrt{90}}}$

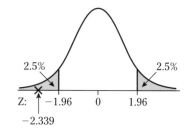

$\qquad\qquad = -2.339... < -1.96$

and $\qquad |z| = 2.339... > 1.96$

Since $|z| > 1.96$, reject H_0 in favour of H_1.

There is evidence, at the 5% level, that the mean number of words in a sentence is not the same as the author's, so you would conclude that the book has been written by a different author.

(ii) A Type I error is made when you say that it is **not** the same author when in fact it **is** the same author.

The probability of making a Type I error in part (i) is the probability that you reject H_0 when $\mu = 21.2$,

so P(Type I error) = 5%

This probability is the same as the significance level of the test.

Example 6.10

Over a long period of time it is found that the time spent at cash withdrawal points follows a normal distribution with mean 2.1 minutes and standard deviation 0.9 minutes. A new system is tried out, to speed up the procedure. The null hypothesis is that the mean time spent is the same under the new system as previously. It is decided to reject the null hypothesis and accept that the new system is quicker if the mean withdrawal time from a random sample of 20 cash withdrawals is less than 1.7 minutes. Assume that, for the new system, the standard deviation is still 0.9 minutes, and the time spent follows a normal distribution.

(i)   Calculate the probability of a Type I error.

(ii) If the mean withdrawal time under the new system is actually 1.5 minutes, calculate the probability of a Type II error.

Cambridge Paper 7 Q5 J03

X is the time, in minutes, spent at the cash withdrawal point, where $X \sim N(\mu, 0.9^2)$

$H_0: \mu = 2.1$ (The mean time is unchanged)

$H_1: \mu < 2.1$ (The mean time has decreased)

If H_0 is true then, for samples of size 20,

$$\bar{X} \sim N\left(2.1, \frac{0.9^2}{20}\right) \qquad \mu = 2.1, \frac{\sigma}{\sqrt{n}} = \frac{0.9}{\sqrt{20}}$$

(i) P(Type I error)

$= $ P(rejecting H_0 when in fact H_0 is true)

$= $ P($\bar{X} < 1.7$ when $\mu = 2.1$)

$= P\left(Z < \dfrac{1.7 - 2.1}{\dfrac{0.9}{\sqrt{20}}}\right)$

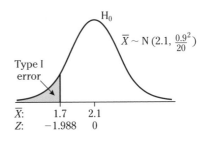

$= $ P($Z < -1.988$)   You need to round to 3 d.p. to use the normal table.

$= 1 - \Phi(1.988)$

$= 1 - 0.9766$

$= 0.0234$

(ii) P(Type II error) = P(accepting H_0 when in fact H_1 is true)

Now you accept H_0 when $\bar{x} \geqslant 1.7$,

so P(Type II error)

$$= P(\bar{X} \geqslant 1.7 \text{ when } \mu = 1.5)$$

$$= P\left(Z > \frac{1.7 - 1.5}{\frac{0.9}{\sqrt{20}}}\right)$$

$$= P(Z > 0.994)$$

$$= 1 - \Phi(1.994)$$

$$= 1 - 0.8399$$

$$= 0.1601$$

$$= 0.160 \text{ (3 s.f.)}$$

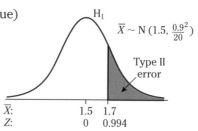

H_1

$\bar{X} \sim N\,(1.5, \frac{0.9^2}{20})$

Type II error

$\bar{X}$: 1.5 1.7
Z: 0 0.994

The following diagram illustrates the situation described above.

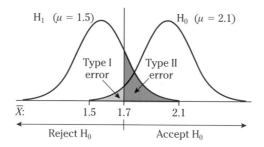

$H_1\ (\mu = 1.5)$ $H_0\ (\mu = 2.1)$

Type I error   Type II error

$\bar{X}$: 1.5 1.7 2.1

Reject H_0 Accept H_0

Exercise 6c

1 A large random sample was taken from a population with mean μ and known variance. The null hypothesis $\mu = 52$ was tested against the alternative hypothesis $\mu \neq 52$ at the 4% significance level. The calculated value of the standardised test statistic was 2.19

   (i) Carry out a hypothesis test for μ based on this result, stating your conclusion clearly.

   (ii) State the probability of making a Type I error.

2 A sample of size 15 is taken from the distribution of X, where $X \sim N(\mu, 4)$

   If the sample mean is greater than 10.72, the null hypothesis $\mu = 10$ is rejected in favour of the alternative hypothesis $\mu > 10$

   (i) Find the probability of making a Type I error.

   (ii) Find the probability of making a Type II error if, in fact, $\mu = 10.5$

3 The recording time of DVDs manufactured by a particular firm may be modelled by a normal distribution with standard deviation 1.8 minutes. The DVDs are advertised as having a recording time of 120 minutes, but the manufacturer claims that they actually have a mean recording time of 123 minutes.

An investigator suspected that the recording time is less than 123 minutes. He selected 36 DVDs at random and calculated the mean recording time. On the basis of the sample mean, he rejected the manufacturer's claim at the 5% level, saying that the mean recording time of the DVDs manufactured by the firm was less than 123 minutes.

   (i) What can be said about the value of the sample mean for this decision to be taken?

   (ii) State the probability of making a Type I error.

4 Boxes of a certain breakfast cereal have contents whose masses, in grams, are normally distributed with mean μ and standard deviation 15. A test of the null hypothesis $\mu = 375$ against the alternative hypothesis $\mu > 375$ is carried out at the 2% significance level using a random sample of 16 boxes.

   (i) Show that the alternative hypothesis is accepted when $\bar{x} > 382.35$, where $\bar{x}$ is the mean mass of the boxes of breakfast cereal in the sample.

   (ii) Given that the actual value of μ is 385, find the probability of making a Type II error.

   (iii) Find the range of values of μ for which the probability of making a Type II error is less than 0.025

   The test is carried out, independently, on two different occasions.

   (iv) Find the probability that at least one Type I error is made.

5 The random variable X is distributed $N(\mu, 3.5^2)$. A test of the null hypothesis $\mu = 15$ against the alternative hypothesis $\mu > 15$ is to be carried out and it is required that the probability of a Type I error should be 0.05

   A random sample of 30 observations on X is taken.

   (i) The critical region is $\bar{X} > c$. Find the value of c.

   The mean of the sample was 16.00

   (ii) Find a 95% confidence interval for μ.

   (iii) Find P(Type II error) for the test in part (i) when $\mu = 17$

6 A random variable has a normal distribution with mean μ and standard deviation 3.

   The null hypothesis $\mu = 20$ is to be tested against the alternative hypothesis $\mu > 20$ using a random sample of size 25. It is decided that the null hypothesis will be rejected if the sample mean is greater than 21.4

   (i) Find the probability of making a Type I error.

   (ii) Find the probability of making a Type II error if in fact $\mu = 21$

7 The random variable X is distributed as $N(\mu, 16)$. A random sample of size 25 is available. The null hypothesis $\mu = 0$ is to be tested against the alternative hypothesis $\mu \neq 0$. The null hypothesis will be accepted if $-1.5 < \bar{x} < 1.5$, where $\bar{x}$ is the value of the sample mean; otherwise the null hypothesis will be rejected.

   (i) Find the probability of a Type I error.

   (ii) Find the probability of a Type II error, if in fact, $\mu = 0.5$

8 When watching games of men's basketball, I have noticed that the players are often tall. I am interested to find out whether or not men who play basketball really are taller than men in general.

   I know that the heights, in metres, of men in general have the distribution $N(1.73, 0.08^2)$. I make the assumption that the heights X, in metres, of male basketball players are also normally distributed, with the same variance as the heights of men in general, but possibly with a larger mean.

   (i) Write down the null and alternative hypotheses under test.

   I propose to base my test on the heights of 8 male basketball players who recently appeared for our local team, and I shall use a 5% level of significance.

   (ii) Write down the distribution of the sample mean, $\bar{X}$, for samples of size 8 drawn from the distribution of X assuming that the null hypothesis is true.

   (iii) Determine the critical region for my test, illustrating your answer with a sketch.

   (iv) Carry out the test, given that the mean height of the 8 players is 1.765 m. Present your conclusions carefully, stating any additional assumption you need to make.

   In fact, the distribution of X is $N(1.80, 0.06^2)$.

   (v) Find the probability that a test based on a random sample of size 8 and using the critical region in part (iii) will lead to the conclusion that male basketball players are *not* taller than men in general.

HYPOTHESIS TEST 2:
TESTING A BINOMIAL PROPORTION *p* WHEN *n* IS LARGE

Consider the situation when independent trials are carried out, each with a probability p of success, where p is constant. If X is the number of successes in n trials, then X follows a binomial distribution, i.e. $X \sim B(n, p)$

In Chapter 5 (page 137) you learned how to carry out a hypothesis test for an unknown binomial proportion p. This involved calculating binomial probabilities which are relatively easy to find when n is small.

When n is large, however, the calculations can become very cumbersome. In such cases it is useful to use the **normal approximation to the binomial distribution** (S1 page 202).

When carrying out the hypothesis test, since the normal distribution is being used, you can work in standardised z-values. However, since the normal distribution is continuous and the binomial is discrete, you will need to use a **continuity correction** which involves amending your test value by adding or subtracting 0.5

Further details are given in the following examples. The stages of the test are the same as in the general procedure outlined on page 164.

The null hypothesis is H_0: $p = p_0$, where p_0 is a specified value.

The alternative hypothesis is H_1: $p < p_0$ (or $p > p_0$ or $p \neq p_0$).

To test a binomial proportion p, the test statistic is X, the number of successes in n trials, where

$$X \sim B(n, p)$$

When **n is large** such that $np > 5$ and $nq > 5$ (where $q = 1 - p$), X is approximately normal and

$$X \sim N(np, npq) \text{ approximately} \qquad \mu = np, \sigma = \sqrt{npq}$$

In standardised form, the test statistic is

$$Z = \frac{X - np}{\sqrt{npq}} \quad \text{where } Z \sim N(0, 1)$$ Note that the value of p is specified in the null hypothesis.

Example 6.11

Caroline is asked to test whether a coin is biased in favour of heads, using a 5% level of significance. She tossed the coin 100 times and obtained 57 heads. Carry out the hypothesis test, stating your null and alternative hypotheses clearly.

X is the number of heads in 100 tosses.

H_0: $p = 0.5$ (the coin is equally likely to show heads or tails)

H_1: $p > 0.5$ (the coin is more likely to show heads than tails)

If H_0 is true, then $X \sim B(100, 0.5)$ $n = 100, p = 0.5, q = 1 - p = 0.5$

Now $np = 100 \times 0.5 = 50$ and $nq = 100 \times 0.5 = 50$. Since $np > 5$ and $nq > 5$ use the normal approximation.

$X \sim N(np, npq)$ with $np = 50$ and $npq = 100 \times 0.5 \times 0.5 = 25$

i.e. $X \sim N(50, 25)$ $\mu = 50$, $\sigma = \sqrt{25} = 5$

Using a one-tail (upper-tail) test at the 5% significance level, the critical z-value is 1.645

So reject H_0 if $z > 1.645$ where $z = \dfrac{x - np}{\sqrt{npq}}$

You are given that 57 heads are obtained.

When standardising the sample value of 57 heads you have to use a continuity correction.

Think of the discrete value of 57 being represented by a rectangle over the continuous interval from 56.5 to 57.5

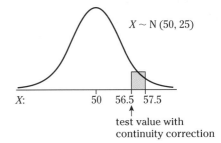

In order to reject H_0, the complete rectangle must lie in the critical region, so the test value is taken as the left hand boundary.

Applying a continuity correction, the test value is $x = 56.5$

So $z = \dfrac{56.5 - 50}{5}$

$= 1.3 < 1.645$

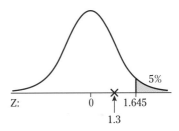

Since $z < 1.645$, do not reject H_0

There is not enough evidence, at the 5% level, to say that the coin is biased in favour of heads.

It is interesting to work out how many heads would need to be obtained to conclude that the coin is biased in favour of heads. This can be done as follows.

The standardised test value z lies in the critical region if $z > 1.645$

If the number of heads is x, then, applying the continuity correction, you need to use $x - 0.5$ when standardising the test value.

So $\dfrac{(x - 0.5) - np}{\sqrt{npq}} > 1.645$

i.e. $\dfrac{(x - 0.5) - 50}{5} > 1.645$

$x > 50 + 0.5 + 1.645 \times 5$

$x > 58.725$

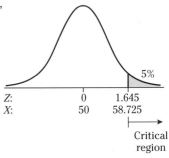

Since x is an integer, the smallest value of x is 59 and the critical region is $X \geqslant 59$. So if Caroline had obtained 59 heads or more when she tossed the coin 100 times, she would have concluded that the coin was biased in favour of heads.

This result is perhaps surprising. Would you have thought that more heads would be needed?

Note: The hypothesis test for a binomial proportion p may be carried out using the distribution of P_s, the sample proportion (see Chapter 4, page 124). However, it is easier to apply the continuity correction using the method described on the previous page.

Example 6.12

A manufacturer claims that a particular brand of seeds has a germination rate of 90%.

To test this claim, 150 randomly selected seeds are planted. It is noted that 124 germinate. Use a suitable approximation to test, at the 1% significance level, whether the manufacturer is overstating the germination rate of the seeds.

X is the number of seeds that germinate in the sample of 150 seeds.

H_0: $p = 0.9$ (the germination rate is 90%)

H_1: $p < 0.9$ (the germination rate is less than 90% and the manufacturer is overstating the rate)

If H_0 is true, then $X \sim B(150, 0.9)$ $n = 150, p = 0.9, q = 0.1$

Now $np = 150 \times 0.9 = 135 > 5$ and $nq = 150 \times 0.1 = 15 > 5$

so the normal approximation can be used.

$X \sim N(np, npq)$ with $np = 135$ and $npq = 150 \times 0.9 \times 0.1 = 13.5$

i.e. $X \sim N(135, 13.5)$ $\mu = 135$, $\sigma = \sqrt{13.5}$

Using a one-tail (lower-tail) test at the 1% significance level, the critical z-value is -2.326.

So reject H_0 if $z < -2.326$ where $z = \dfrac{x - np}{\sqrt{npq}}$

You are given that 124 seeds germinate.

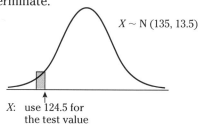

$X \sim N (135, 13.5)$

X: use 124.5 for the test value

Think of the discrete value of 124 being represented by a rectangle over the continuous interval from 123.5 to 124.5.

In order to reject H_0, the complete rectangle must lie in the critical region, so the test value is taken as the right-hand boundary.

Applying a continuity correction, the test value is $x = 124.5$

So $z = \dfrac{124.5 - 135}{\sqrt{13.5}}$ Input $\sqrt{13.5}$ on your calculator. You will lose accuracy if you use a rounded value.

$= -2.857... < -2.326$

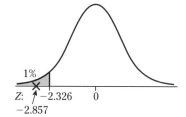

1%

$Z: -2.326$ 0

-2.857

Since $z < -2.326$, reject H_0 in favour of H_1.

There is evidence, at the 1% level, that the germination rate is less than 90%, so the manufacturer is overstating the rate.

Example 6.13

When cars arrive at a certain junction they can turn either right or left. Part of a study of road usage involves deciding between the following alternatives:

- Cars are equally likely to turn right or left
- Cars are more likely to turn right than left.

(i) A hypothesis test is to be carried out, at the 2% significance level, based on a sample of 40 cars.

    (a) State suitable null and alternative hypotheses,

    (b) Find the critical region.

(ii) For the test described in part (i), calculate the probability of making a Type II error when, in fact, 80% of all cars arriving at the junction turn right.

Let X be the number of cars that turn right in a random sample of 40 cars, where $X \sim B(40, p)$.

(i) (a) $H_0: p = 0.5$ (cars are equally likely to turn right or left)

      $H_1: p > 0.5$ (cars are more likely to turn right than left)

  (b) If H_0 is true, then

$$X \sim B(40, 0.5) \qquad n = 40, p = 0.5, q = 0.5$$

Now $np = 40 \times 0.5 = 20 > 5$ and $nq = 40 \times 0.5 = 20 > 5$

Using the normal approximation,

$$X \sim N(np, npq) \text{ with } np = 20, \ npq = 40 \times 0.5 \times 0.5 = 10$$

i.e. $X \sim N(20, 10) \qquad \mu = 20, \ \sigma = \sqrt{10}$

Using a one-tail (upper-tail) test at the 2% level, to find the critical region you need to find the least integer value of c such that $P(X \geqslant c) < 0.02$

Applying a continuity correction, the rejection rule becomes

$$P(X > (c - 0.5)) < 0.02$$

Standardising:

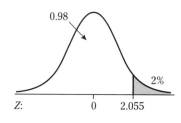

$$P\left(Z > \frac{(c - 0.5) - 20}{\sqrt{10}}\right) < 0.02$$

Using the Normal distribution table,

$$P(Z \leqslant 2.055) = 0.98, \text{ so } P(Z > 2.055) = 0.02$$

Therefore $\dfrac{(c - 0.5) - 20}{\sqrt{10}} > 2.055$

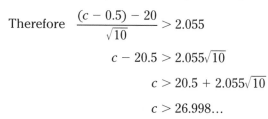

$$c - 20.5 > 2.055\sqrt{10}$$

$$c > 20.5 + 2.055\sqrt{10}$$

$$c > 26.998\ldots$$

Since c is an integer, $c = 27$. So the critical region is $X \geqslant 27$

(ii) You are given H_1: $p = 0.8$

When $p = 0.8$,

$$np = 40 \times 0.8 = 32 > 5 \text{ and } nq = 40 \times 0.2 = 8 > 5$$

Use a normal approximation, where

$$X \sim N(np, npq) \text{ with } np = 40, npq = 40 \times 0.8 \times 0.2 = 6.4,$$

i.e. $X \sim N(32, 6.4)$ $\mu = 32, \ \sigma = \sqrt{6.4}$

$P(\text{Type II error}) = P(\text{accept } H_0 \text{ when } H_1 \text{ is true})$

$$= P(\text{accept } H_0 \text{ when } p = 0.8)$$

In the test in part (i), H_0 is accepted if $X < 27 \rightarrow X < 26.5$ (continuity correction)

$P(\text{Type II error}) = P(X < 26.5 \text{ when } X \sim N(32, 6.4))$

$$= P\left(Z < \frac{26.5 - 32}{\sqrt{6.4}}\right)$$

$$= P(Z < -2.174)$$

$$= 1 - 0.9852$$

$$= 0.0148$$

So, $P(\text{Type II error}) \approx 1.5\%$

P(Type II error)

| X: | 26.5 | 32 |
| Z: | −2.174 | 0 |

Exercise 6d

1 In the following, $X \sim B(n, p)$ with n as shown, p is unknown and x is the number of successes in the sample. Test the hypotheses stated at the level of significance indicated.

	Distribution of X	Sample value x	Hypotheses		Significance level
(i)	$X \sim B(50, p)$	45	H_0: $p = 0.8$	H_1: $p > 0.8$	5%
(ii)	$X \sim B(60, p)$	42	H_0: $p = 0.55$	H_1: $p > 0.55$	10%
(iii)	$X \sim B(120, p)$	21	H_0: $p = \frac{1}{4}$	H_1: $p < \frac{1}{4}$	5%
(iv)	$X \sim B(300, p)$	213	H_0: $p = 0.65$	H_1: $p \neq 0.65$	1%
(v)	$X \sim B(90, p)$	56	H_0: $p = 0.76$	H_1: $p < 0.76$	1%

2 In an investigation into the ownership of cell phones among schoolchildren in a particular city, 200 randomly chosen schoolchildren from the city were interviewed and it was found that 142 owned a cell phone.

Use a suitable approximation to test, at the 5% level of significance, the null hypothesis that 65% of schoolchildren in the city own a cell phone against the alternative hypothesis that more than 65% own a cell phone.

3 A manufacturer claims that as many as 8 out 10 dogs prefer its brand of dog food to any other, but a consumer group believes that the proportion is lower. In a random sample of

120 dogs, it was found that 85 appeared to prefer that particular brand of dog food.

Use a suitable approximation to test, at the 5% significance level, whether the manufacturer's claim is supported.

4 In a survey it was found that 30% of voters supported a particular political party. A month later a party representative claimed that the popularity of the party had increased. In a random sample of 100 voters, 38 supported the party.

Use a suitable approximation to test, at the 3% significance level, whether support for the party has increased.

5 A large college claims that it admits equal numbers of men and women. In a random sample of 500 students at the college there were 267 males.

Use a suitable approximation to test, at the 5% significance level, whether this provides evidence that the college population is **not** equally divided between males and females.

6 A theory predicts that the probability of an event is 0.4

The theory is tested experimentally and in 400 independent trials the event occurred 140 times.

Use a suitable approximation to test, at the 1% significance level, whether the proportion is less than that predicted by the theory.

7 It is claimed that the proportion of defective items produced by a particular machine is 0.1

A random sample of 100 items is inspected and found to contain 15 defective items.

Use a suitable approximation to test, at the 5% significance level, whether the proportion of defective items is 0.1

8 (i) A gardener sows 150 Special cabbage seeds and knows that the germination rate is 75%

By using a suitable approximation find the probability that:

(a) more than 122 seeds germinate

(b) fewer than 106 seeds germinate.

(ii) The gardener also sows 120 Everyday cabbage seeds and finds that 81 germinate. Use a suitable approximation to test, at the 4% significance level, whether the Everyday seeds have a germination rate less than 75%

9 A government report states that a third of teenagers in a certain region belong to a youth organisation. A survey, conducted among a random sample of 1000 teenagers from the region revealed that 370 belonged to a youth organisation. Does this provide evidence, at the 2% level, that more than a third of teenagers in the region belong to a youth organisation?

10 A questionnaire was sent to a large number of teachers, asking for their opinions about a proposal to alter an examination syllabus. Of the 180 replies received, 134 were in favour of the proposal.

(i) Stating a necessary assumption, use a suitable approximation to test, at the 5% significance level, the null hypothesis that the population proportion in favour of the proposal is 0.7 against the alternative hypothesis that it is more than 0.7

(ii) Assuming the assumption in part (i) is valid, find a 95% confidence interval for the population proportion in favour of the proposal.

11 After carrying out a survey, a market research company asserted that 75% of TV viewers watched a certain programme. Another company interviewed 75 viewers and found that 51 had watched the programme and 24 had not. Use a suitable approximation to test, at the 5% significance level, whether the first market research company's figure of 75% was incorrect.

12 The manager of a large supermarket wishes to judge the effect on the customers of a new layout. On the day that the layout was introduced the first 200 customers in the store were asked whether or not they approved of the new layout.

(i) Comment on the manner in which the sample was chosen, and suggest a way of obtaining a more suitable sample.

Out of a suitably chosen sample of 200 customers, 148 approved of the new layout. The manager claims that 80% of customers approve of the new layout.

(ii) Use a suitable approximation to show that the data provide evidence at the 2% significance level that the population percentage is less than 80%

13 The random variable X can be modelled by a binomial distribution with parameters n and p, where $n = 200$ and p is unknown.

(i) A hypothesis test is carried out to test the null hypothesis $p = 0.4$ against the alternative hypothesis $p < 0.4$

The sample value is m. Use a suitable approximation to find the greatest value of m for which the null hypothesis is rejected at the 5% significance level.

(ii) For the test described in part (i), find the probability of making a Type II error if, in fact, $p = 0.3$

14 The process of manufacturing a certain kind of dinner plate results in a proportion 0.13 of faulty plates. An alteration is made to the process which is intended to reduce the proportion of faulty plates.

(i) (a) State suitable null and alternative hypotheses for a hypothesis test to test the effectiveness of the alteration.

In order to carry out the test, a member of the quality control department decides to count the number of faulty plates in a random sample of 2500. If 290 or fewer faulty plates are found then she will accept that the alteration does result in a reduction in the proportion of faulty plates.

(b) Determine the significance level of this test, using a suitable normal approximation.

(ii) Determine the probability of making a Type II error in the test in part (i) if the alteration does in fact result in a decrease in the proportion of faulty plates to 0.11

HYPOTHESIS TEST 3: TESTING A POISSON MEAN WHEN λ IS LARGE

In Chapter 5 (page 145) you learned how to carry out a hypothesis test for λ, the mean of a Poisson distribution. This involved calculating Poisson probabilities by direct evaluation.

If X has a Poisson distribution with **large** λ, then a **normal approximation** to the Poisson distribution can be used (see page 20).

The null hypothesis is $H_0: \lambda = \lambda_0$, where λ_0 is a specified value.

The alternative hypothesis is $H_1: \lambda < \lambda_0$ (or $\lambda > \lambda_0$ or $\lambda \neq \lambda_0$).

To test a Poisson mean λ, the test statistic is X, where

$$X \sim Po(\lambda)$$

When λ is **large** ($\lambda > 15$)

$$X \sim N(\lambda, \lambda) \text{ approximately} \qquad \mu = \lambda \text{ and } \sigma = \sqrt{\lambda}$$

In standardised form, the test statistic is

$$Z = \frac{X - \lambda}{\sqrt{\lambda}} \qquad \text{where } Z \sim N(0, 1)$$

Note that the value of λ is specified in the null hypothesis.

When carrying out the hypothesis test, since the normal distribution is being used, you can work in standardised z-values. However, as with the binomial tests, since the normal distribution is continuous and the Poisson distribution is discrete, you will need to use a **continuity correction**, as illustrated in the following examples.

The stages of the test are the same as in the general procedure outlined on page 164.

Example 6.14

The number of hits per day, X, on a property website can be modelled by a Poisson distribution with mean 75. On a randomly chosen day following an advertising campaign there were 89 hits. Test at the 3% significance level whether the mean number of hits per day has increased.

X is the number of hits in a day.

> H_0: $\lambda = 75$ (the mean number of hits is unchanged)
>
> H_1: $\lambda > 75$ (the mean number of hits has increased)

If H_0 is true, then $X \sim \text{Po}(75)$.

Since λ is large,

> $X \sim N(75, 75)$ approximately, with $\mu = 75$ and $\sigma = \sqrt{75}$.

Use a one-tailed (upper-tail) test at the 3 % significance level.

The critical z-value is $\Phi^{-1}(0.97) = 1.881$.

Reject H_0 if $z > 1.881$, where $z = \dfrac{x - 75}{\sqrt{75}}$.

Following the advertising campaign, the number of hits is 89. Applying the continuity correction, the test value is $x = 88.5$

So $\quad z = \dfrac{88.5 - 75}{\sqrt{75}}$

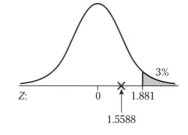

$\qquad = 1.5588\ldots < 1.881$

Since $z < 1.881$, do not reject H_0.

There is not enough evidence, at the 3% level, that the mean number of hits per day has increased following the advertising campaign.

Example 6.15

Flies stick to wet paint at random points. The average number of flies is 2 per square metre. A wall with area $22\,\text{m}^2$ is painted with a new type of paint which the manufacturer claims is fly-repellent. It is found that 27 flies stick to this wall. Use a suitable approximation to test the manufacturer's claim at the 1% significance level. Take the null hypothesis to be $\mu = 44$, where μ is the population mean.

Cambridge Paper 7 Q3 N05

X is the number of flies on $22\,\text{m}^2$ of wall, where X follows a Poisson distribution.

Take care with the unit interval. If the average number of flies in a square metre is 2, the average number in $22\,m^2$ is 44.

> H_0: $\mu = 44$ (The average number of flies is unchanged)
>
> H_1: $\mu < 44$ (The average number of flies sticking to the wet paint has decreased)

If H_0 is true, then $X \sim \text{Po}(44)$

Since μ is large,

$X \sim \text{N}(44, 44)$ with $\mu = 44$ and $\sigma = \sqrt{44}$

Use a one-tail (lower-tail) test at the 1% significance level.

The critical z-value is -2.326, so reject H_0 if $z < -2.326$, where $z = \dfrac{x - 44}{\sqrt{44}}$

With the new paint there were 27 flies.

Applying a continuity correction, the test value is $x = 27.5$

So $z = \dfrac{27.5 - 44}{\sqrt{44}}$

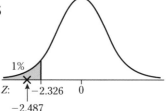

$= -2.487... < -2.326$

Since $z < -2.326$, reject H_0 in favour of H_1.

There is evidence, at the 1% level, that the mean
number of flies has decreased, supporting the manufacturer's claim.

Example 6.16

The number of cases, X, of a particular disease treated each year at a hospital follows a
Poisson distribution with mean 56.

Following a campaign to alert people to preventative measures, it is believed that the mean
number of cases treated in a year will decrease. A hypothesis test is to be carried out to test
the null hypothesis $\lambda = 56$ against the alternative hypothesis $\lambda < 56$. It is decided to reject
the null hypothesis if $X \leqslant 40$

(i)   Find the probability of making a Type I error.

(ii) Find the probability of making a Type II error if $\lambda = 45$

X is the number of cases treated in a year, where

$H_0: \lambda = 56$ (The mean number of cases is unchanged)

$H_1: \lambda < 56$ (The mean number of cases has decreased)

If H_0 is true, then $X \sim \text{Po}(56)$

Since λ is large,

$X \sim \text{N}(56, 56)$ with $\mu = 56$ and $\sigma = \sqrt{56}$

A one-tail (lower-tail) test is to be carried out and H_0 is rejected if $X \leqslant 40$

P(Type I error) = P(reject H_0 when H_0 is true)

$= \text{P}(X \leqslant 40 \text{ when } \lambda = 56)$

$= \text{P}\left(Z < \dfrac{40.5 - 56}{\sqrt{56}}\right)$ continuity correction

$= \text{P}(Z < -2.071)$

$= 1 - \Phi(2.071)$

$= 1 - 0.9808$

$= 0.0192$

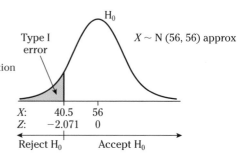

(ii) You are given that $\lambda = 45$

Since λ is large, $X \sim N(45, 45)$

P(Type II error) = P(accept H_0 when H_1 is true)

$$= P(X > 40 \text{ when } \lambda = 45)$$

$$= P\left(Z > \frac{40.5 - 45}{\sqrt{45}}\right) \quad \text{continuity correction}$$

$$= P(Z > -0.671)$$

$$= \Phi(0.671)$$

$$= 0.7489$$

$$= 0.749 \text{ (3 s.f.)}$$

Exercise 6e

1 In the following, $X \sim Po(\lambda)$, where λ is unknown and x is the number of occurrences in the sample. Test the hypotheses stated at the level of significance indicated.

	Sample value x	Hypotheses	Significance level
(i)	71	$H_0: \lambda = 49$ $H_1: \lambda > 49$	5%
(ii)	113	$H_0: \lambda = 130$ $H_1: \lambda < 130$	10%
(iii)	13	$H_0: \lambda = 25$ $H_1: \lambda \neq 25$	1%
(iv)	76	$H_0: \lambda = 68$ $H_1: \lambda > 68$	6%
(v)	227	$H_0: \lambda = 260$ $H_1: \lambda < 260$	2.5%

2 A traffic survey shows that vehicles pass a particular census point between 11.00am and 11.30am at an average rate of 3 per minute.

When a nearby road is closed for repairs, it is claimed that the number of vehicles passing the census point has increased. Local residents conduct a survey and find that 110 vehicles pass the census point between 11.00am and 11.30am on a randomly chosen day.

Using a suitable approximation, carry out a hypothesis test at the 5% significance level to test whether the mean number of vehicles passing the census point has increased.

3 Flaws in a certain brand of tape occur independently and at random at an average rate of 2 per 100 m. A new process is introduced in the manufacture of the tape and inspection of a 2 km length reveals 24 flaws.

Using a suitable approximation, test at the 1% significance level whether the mean number of flaws in the tape has decreased.

4 The number of vacuum cleaners sold in a certain shop may be modelled by a Poisson distribution with parameter 3 per day. The shopkeeper placed an advertisement in the local paper and during the following 6 days 24 vacuum cleaners were sold.

Using a suitable approximation, test at the 6% significance level whether there has been an increase in the mean number of vacuum cleaners sold.

5 When traffic is flowing freely, the number of vehicles arriving at the toll booths on a bridge follows a Poisson distribution. It is believed that the mean number of vehicles arriving per minute is 4.2

To test this, the flow arrivals were monitored during 30 min when traffic was flowing freely. During this time, 96 vehicles arrived. Test, at the 5% significance level, whether the mean number of vehicles arriving per minute is 4.2

6 The random variable X follows a Poisson distribution with mean λ.

A hypothesis test is carried out to test the null hypothesis $\lambda = 34.2$ against the alternative hypothesis $\lambda > 34.2$

A random observation of X is made and it is decided to reject the null hypothesis if $X \geqslant 42$

(i) Find the probability of making a Type I error.

(ii) Find the probability of making a Type II error if $\lambda = 39.6$

7 Over a period of time it is found that, on the day that a new window display is installed at a certain shop, the number of people who stop to look at the display follows a Poisson distribution with a mean of 1 person per minute.

A new window dresser is employed, and during a randomly chosen hour following the installation of the new display, 80 people stop to look at it. Test, at the 1% significance level, whether there is evidence that the mean number of people stopping to look at the window display has increased.

Summary

Steps in a hypothesis test (*z*-tests)

- Define the variable
- State H_0 and H_1, then give the distribution of the test statistic assuming H_0 is true
- State the rejection rule – this is usually given in terms of the critical *z*-value
- Find whether the test value lies in the critical region – this usually involves finding a *z*-value and comparing it with the critical *z*-value
- Make your conclusion in statistical terms by saying whether H_0 is rejected or not and then **relate it to the situation**

Critical region

If the test value is in the critical region, reject H_0.
If the test value is not in the critical region, do not reject H_0.

Probability statements for rejection rule

Significance level	One-tail (lower-tail) $H_0: \mu = \mu_0$ $H_1: \mu < \mu_0$	One-tail (upper-tail) $H_0: \mu = \mu_0$ $H_1: \mu > \mu_0$	Two-tail $H_0: \mu = \mu_0$ $H_1: \mu \neq \mu_0$
10%	Reject H_0 if $z < -1.282$	Reject H_0 if $z > 1.282$	Reject H_0 if $z > 1.645$ or $z < -1.645$ i.e. if $|z| > 1.645$
5%	Reject H_0 if $z < -1.645$	Reject H_0 if $z > 1.645$	Reject H_0 if $z > 1.96$ or $z < -1.96$ i.e. if $|z| > 1.96$
1%	Reject H_0 if $z < -2.326$	Reject H_0 if $z > 2.326$	Reject H_0 if $z > 2.576$ or $z < -2.576$ i.e. if $|z| > 2.576$

Test 1: testing a population mean μ based on a sample mean $\bar{x}$

Case 1

Normal population X with known variance σ^2 based on a sample of any size:

Test statistic is $\bar{X}$, where $\bar{X} \sim N\left(\mu, \dfrac{\sigma^2}{n}\right)$

Standardised test statistic is $Z = \dfrac{\bar{X} - \mu}{\dfrac{\sigma}{\sqrt{n}}}$, where $Z \sim N(0, 1)$

The value of μ is specified in the null hypothesis.

Case 2

Non-normal population X with known variance σ^2, based on the mean of a **large** sample, size $n \geqslant 30$:

Test statistic is $\overline{X}$, where $\overline{X} \sim N\left(\mu, \dfrac{\sigma^2}{n}\right)$ approximately (by the Central Limit theorem).

Standardised test statistic is $Z = \dfrac{\overline{X} - \mu}{\dfrac{\sigma}{\sqrt{n}}}$ where $Z \sim N(0, 1)$

Case 3

Population X with unknown variance σ^2, based on the mean of a **large** sample, size $n \geqslant 30$:

Use s^2, the unbiased estimate of the population variance, where $s^2 = \dfrac{1}{n - 1}\left(\Sigma x^2 - \dfrac{(\Sigma x)^2}{n}\right)$

The test statistic is $\overline{X}$, where $\overline{X} \sim N\left(\mu, \dfrac{s^2}{n}\right)$ approximately.

Standardised test statistic is $Z = \dfrac{\overline{X} - \mu}{\dfrac{s}{\sqrt{n}}}$ where $Z \sim N(0, 1)$

Test 2: testing a binomial population proportion p when n is large

X is the number of successes in n trials, where $X \sim B(n, p)$ The value of p is specified in the null hypothesis.

When n is large such that $np > 5$ and $nq > 5$, then $X \sim N(np, npq)$ approximately.

Standardised test statistic is $Z = \dfrac{X - np}{\sqrt{npq}}$ where $Z \sim N(0, 1)$ Remember to use a continuity correction (± 0.5).

Test 3: testing a Poisson mean λ when λ is large

X is the number of occurrences in a given interval, where $X \sim Po(\lambda)$ The value of λ is specified in the null hypothesis.

When λ is large ($\lambda > 15$), then $X \sim N(\lambda, \lambda)$ approximately.

Standardised test statistic is $Z = \dfrac{X - \lambda}{\sqrt{\lambda}}$ where $Z \sim N(0, 1)$ Remember to use a continuity correction (± 0.5).

Type I and Type II errors

A Type I error is made when the null hypothesis is rejected but it is in fact true.

A Type II error is made when the null hypothesis is accepted but it is in fact false.

$P(\text{Type I error}) = P(\text{reject } H_0 \text{ when } H_0 \text{ is true})$

$P(\text{Type II error}) = P(\text{accept } H_0 \text{ when } H_1 \text{ is true})$

To find the probability of making a Type II error you must be given a specific value for the alternative hypothesis H_1.

Mixed Exercise 6

1 A machine has produced nails over a long period of time, where the length in millimetres was distributed as N(22.0, 0.19). It is believed that recently the mean length has changed. To test this belief a random sample of 8 nails is taken and the mean length is found to be 21.7 mm. Carry out a hypothesis test at the 5% significance level to test whether the population mean has changed, assuming that the variance remains the same.

Cambridge Paper 7 Q3 J07

2 The time taken for Samuel to drive home from work is distributed with mean 46 minutes. Samuel discovers a different route and decides to test at the 5% level whether the mean time has changed. He tries this route on a large number of different days chosen randomly and calculates the mean time.

(i) State the null and alternative hypotheses for this test.

(ii) Samuel calculates the value of his test statistic z to be -1.729. What conclusion can he draw?

Cambridge Paper 7 Q1 N06

3 A factory produces ropes. The breaking strengths of the ropes are normally distributed with standard deviation 4.7 kg. Although the ropes are advertised as having a mean breaking strength of 120 kg, the manufacturer claims that the mean breaking strength is greater than 120 kg. A random sample of 16 ropes had a mean breaking strength of 123 kg. Does this provide evidence, at the 2.5% level, to support the manufacturer's claim?

4 The distance driven in a week by a long-distance lorry driver is a normally distributed random variable with mean 1850 km and standard deviation 117 km.

(i) Find the probability that in a random sample of 26 weeks his average distance driven per week is more than 1800 km.

(ii) New driving regulations are introduced and in a random sample of 26 weeks after their introduction the lorry driver drives a total of 47 658 km. Assuming the standard deviation remains unchanged, test at the 10% level whether his mean weekly driving distance has changed.

Cambridge Paper 7 Q5 N03

5 A clinic monitors the amount, X milligrams per litre, of a certain chemical in the blood stream of patients. For patients who are taking drug A, it has been found that the mean value of X is 0.336. A random sample of 100 patients taking a new drug, B, was selected and the values of X were found.

The results are summarised below.

$$n = 100, \Sigma x = 43.5 \text{ and } \Sigma x^2 = 31.56$$

(i) Test at the 1% significance level whether the mean amount of the chemical in the blood stream of patients taking drug B is different from that of patients taking drug A.

(ii) For the test to be valid, is it necessary to assume a normal distribution for the amount of chemical in the blood stream of patients taking drug B? Justify your answer.

Cambridge Paper 7 Q6 N10(73)

6 The proportion of patients who suffer an allergic reaction to a certain drug used to treat a particular medical condition is assumed to be 0.045

When 400 patients were treated, 25 suffered an allergic reaction. Using a suitable approximation, test at the 5% significance level whether the quoted figure of 0.045 is an underestimate.

7 A study of the annual rainfall, x centimetres, over the last 20 years for a small town gave the following results:

$$\Sigma x = 1325, \Sigma x^2 = 90\,316$$

(i) Find unbiased estimates of the mean and the variance of the annual rainfall for this town.

Archive records show that the annual rainfall for this town, prior to this period, had mean 62.50 and standard deviation 11.45.

(ii) Assuming that the standard deviation remains unchanged, test at the 5% level of significance whether there is evidence of an increase in mean annual rainfall over the last 20 years. State your hypotheses clearly.

8 When a drawing pin is dropped on to the floor, the probability that it lands point up is p.

   (i) A teacher drops a drawing pin 900 times and observes that it lands point up 315 times. Using a suitable approximation, test, at the 1% significance level, the null hypothesis that $p = 0.4$ against the alternative hypothesis $p < 0.4$

   (ii) A student drops a drawing pin 600 times and observes that it lands point up 251 times. Using the student's results, find a symmetric 95% confidence interval for p.

   (iii) As part of a statistics investigation, 1500 students carry out similar experiments and they each calculate (correctly) their own symmetric 95% confidence interval for p. Find the expected number of these intervals that do not contain the true value of p.

9 The lectures in a mathematics department are scheduled to last 54 minutes, and the times of individual lectures may be assumed to have a normal distribution with mean μ minutes and standard deviation 3.1 minutes. One of the students commented that, on average, the lectures seemed too short. To investigate this, the times for a random sample of 10 lectures were used to test the null hypothesis $\mu = 54$ against the alternative hypothesis $\mu < 54$ at the 10% significance level.

   (i) Show that the null hypothesis is rejected in favour of the alternative hypothesis if $\bar{x} < 52.74$, where $\bar{x}$ minutes is the sample mean.

   (ii) Find the probability of a Type II error given that the actual mean length of lectures is 51.5 minutes.

   Cambridge Paper 7 Q5 J04

10 The number of cars caught speeding on a certain length of motorway is 7.2 per day, on average. Speed cameras are introduced and the results shown in the following table are those from a random selection of 40 days after this.

Number of cars caught speeding	4	5	6	7	8	9	10
Number of days	5	7	8	10	5	2	3

   (i) Calculate unbiased estimates of the population mean and variance of the number of cars per day caught speeding after the speed cameras were introduced.

   (ii) Taking the null hypothesis H_0 to be $\mu = 7.2$, test at the 5% level whether there is evidence that the introduction of speed cameras has resulted in a reduction in the number of cars caught speeding.

   (iii) State what is meant by a Type I error in words relating to the context of the test in part (ii). Without further calculation, illustrate on a suitable diagram the region representing the probability of this Type I error.

   Cambridge Paper 7 Q7 J06

11 People who diet can expect to lose an average of 3 kg in a month. In a book, the authors claim that people who follow a new diet will lose an average of more than 3 kg in a month. The weight losses of the 180 people in a random sample who had followed the new diet for a month were noted. The mean was 3.3 kg and the standard deviation was 2.8 kg.

   (i) Test the authors' claim at the 5% significance level, stating your null and alternative hypotheses.

   (ii) State what is meant by a Type II error in words relating to the context of the test in part (i).

   Cambridge Paper 7 Q4 J08

12 The number of goals scored by a certain football team last season followed a Poisson distribution with mean 1.2 goals per match. This season there are several new players and in a random sample of 20 games the total number of goals scored was 31. Test, at the 10% significance level, whether the mean number of goals has increased this season.

13 Photographers often need to take many photographs of families until they find a photograph which everyone in the family likes. The number of photographs taken until obtaining one which everybody likes has mean 15.2. A new photographer claims that she can obtain a photograph which everybody likes with fewer photographs

taken. To test at the 10% level of significance whether this claim is justified, the numbers of photographs, x, taken by the new photographer with a random sample of 60 families are recorded. The results are summarised by $\Sigma x = 890$ and $\Sigma x^2 = 13\,780$.

(i) Calculate unbiased estimates of the population mean and variance of the number of photographs taken by the new photographer.

(ii) State null and alternative hypotheses for the test, and state also the probability that the test results in a Type I error. Say what a Type I error means in the context of the question.

(iii) Carry out the test.

Cambridge Paper 7 Q6 N09(71)

14 The random variable X has a normal distribution with mean μ (unknown) and variance σ^2 (known). To test the null hypothesis H_0: $\mu = \mu_0$ a random sample of n observations of X is taken and the sample mean, $\bar{x}$, is calculated.

Find, in terms of μ_0, σ and n, the set of values of $\bar{x}$ that will result in each of the following:

(i) H_0 being rejected in favour of H_1: $\mu \neq \mu_0$ at the 5% significance level.

(ii) H_0 not being rejected in favour of H_1: $\mu < \mu_0$ at the 1% significance level.

Normal distribution tables

The normal distribution function

If Z has a normal distribution with mean 0 and variance 1 then, for each value of z, the table gives the value of $\Phi(z)$, where

$$\Phi(z) = P(Z \leq z).$$

For negative values of z use $\Phi(-z) = 1 - \Phi(z)$.

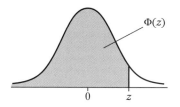

z	0	1	2	3	4	5	6	7	8	9	1	2	3	4	5	6	7	8	9
															ADD				
0.0	0.5000	0.5040	0.5080	0.5120	0.5160	0.5199	0.5239	0.5279	0.5319	0.5359	4	8	12	16	20	24	28	32	36
0.1	0.5398	0.5438	0.5478	0.5517	0.5557	0.5596	0.5636	0.5675	0.5714	0.5753	4	8	12	16	20	24	28	32	36
0.2	0.5793	0.5832	0.5871	0.5910	0.5948	0.5987	0.6026	0.6064	0.6103	0.6141	4	8	12	15	19	23	27	31	35
0.3	0.6179	0.6217	0.6255	0.6293	0.6331	0.6368	0.6406	0.6443	0.6480	0.6517	4	7	11	15	19	22	26	30	34
0.4	0.6554	0.6591	0.6628	0.6664	0.6700	0.6736	0.6772	0.6808	0.6844	0.6879	4	7	11	14	18	22	25	29	32
0.5	0.6915	0.6950	0.6985	0.7019	0.7054	0.7088	0.7123	0.7157	0.7190	0.7224	3	7	10	14	17	20	24	27	31
0.6	0.7257	0.7291	0.7324	0.7357	0.7389	0.7422	0.7454	0.7486	0.7517	0.7549	3	7	10	13	16	19	23	26	29
0.7	0.7580	0.7611	0.7642	0.7673	0.7704	0.7734	0.7764	0.7794	0.7823	0.7852	3	6	9	12	15	18	21	24	27
0.8	0.7881	0.7910	0.7939	0.7967	0.7995	0.8023	0.8051	0.8078	0.8106	0.8133	3	5	8	11	14	16	19	22	25
0.9	0.8159	0.8186	0.8212	0.8238	0.8264	0.8289	0.8315	0.8340	0.8365	0.8389	3	5	8	10	13	15	18	20	23
1.0	0.8413	0.8438	0.8461	0.8485	0.8508	0.8531	0.8554	0.8577	0.8599	0.8621	2	5	7	9	12	14	16	19	21
1.1	0.8643	0.8665	0.8686	0.8708	0.8729	0.8749	0.8770	0.8790	0.8810	0.8830	2	4	6	8	10	12	14	16	18
1.2	0.8849	0.8869	0.8888	0.8907	0.8925	0.8944	0.8962	0.8980	0.8997	0.9015	2	4	6	7	9	11	13	15	17
1.3	0.9032	0.9049	0.9066	0.9082	0.9099	0.9115	0.9131	0.9147	0.9162	0.9177	2	3	5	6	8	10	11	13	14
1.4	0.9192	0.9207	0.9222	0.9236	0.9251	0.9265	0.9279	0.9292	0.9306	0.9319	1	3	4	6	7	8	10	11	13
1.5	0.9332	0.9345	0.9357	0.9370	0.9382	0.9394	0.9406	0.9418	0.9429	0.9441	1	2	4	5	6	7	8	10	11
1.6	0.9452	0.9463	0.9474	0.9484	0.9495	0.9505	0.9515	0.9525	0.9535	0.9545	1	2	3	4	5	6	7	8	9
1.7	0.9554	0.9564	0.9573	0.9582	0.9591	0.9599	0.9608	0.9616	0.9625	0.9633	1	2	3	4	4	5	6	7	8
1.8	0.9641	0.9649	0.9656	0.9664	0.9671	0.9678	0.9686	0.9693	0.9699	0.9706	1	1	2	3	4	4	5	6	6
1.9	0.9713	0.9719	0.9726	0.9732	0.9738	0.9744	0.9750	0.9756	0.9761	0.9767	1	1	2	2	3	4	4	5	5
2.0	0.9772	0.9778	0.9783	0.9788	0.9793	0.9798	0.9803	0.9808	0.9812	0.9817	0	1	1	2	2	3	3	4	4
2.1	0.9821	0.9826	0.9830	0.9834	0.9838	0.9842	0.9846	0.9850	0.9854	0.9857	0	1	1	2	2	2	3	3	4
2.2	0.9861	0.9864	0.9868	0.9871	0.9875	0.9878	0.9881	0.9884	0.9887	0.9890	0	1	1	1	2	2	2	3	3
2.3	0.9893	0.9896	0.9898	0.9901	0.9904	0.9906	0.9909	0.9911	0.9913	0.9916	0	1	1	1	1	2	2	2	2
2.4	0.9918	0.9920	0.9922	0.9925	0.9927	0.9929	0.9931	0.9932	0.9934	0.9936	0	0	1	1	1	1	1	2	2
2.5	0.9938	0.9940	0.9941	0.9943	0.9945	0.9946	0.9948	0.9949	0.9951	0.9952	0	0	0	1	1	1	1	1	1
2.6	0.9953	0.9955	0.9956	0.9957	0.9959	0.9960	0.9961	0.9962	0.9963	0.9964	0	0	0	0	1	1	1	1	1
2.7	0.9965	0.9966	0.9967	0.9968	0.9969	0.9970	0.9971	0.9972	0.9973	0.9974	0	0	0	0	0	1	1	1	1
2.8	0.9974	0.9975	0.9976	0.9977	0.9977	0.9978	0.9979	0.9979	0.9980	0.9981	0	0	0	0	0	0	0	1	1
2.9	0.9981	0.9982	0.9982	0.9983	0.9984	0.9984	0.9985	0.9985	0.9986	0.9986	0	0	0	0	0	0	0	0	0

Critical values for the normal distribution

If Z has a normal distribution with mean 0 and variance 1 then, for each value of p, the table gives the value of z such that

$$P(Z \leq z) = p.$$

p	0.75	0.90	0.95	0.975	0.99	0.995	0.9975	0.999	0.9995
z	0.674	1.282	1.645	1.960	2.326	2.576	2.807	3.090	3.291

List of formulae provided in the exam

Probability and statistics

Summary statistics

For ungrouped data:

$$\bar{x} = \frac{\Sigma x}{n}, \qquad \text{standard deviation} = \sqrt{\frac{\Sigma(x - \bar{x})^2}{n}} = \sqrt{\frac{\Sigma x^2}{n} - \bar{x}^2}$$

For grouped data:

$$\bar{x} = \frac{\Sigma xf}{\Sigma f}, \qquad \text{standard deviation} = \sqrt{\frac{\Sigma(x - \bar{x})^2 f}{\Sigma f}} = \sqrt{\frac{\Sigma x^2 f}{\Sigma f} - \bar{x}^2}$$

Discrete random variables

$$E(X) = \Sigma xp$$
$$\text{Var}(X) = \Sigma x^2 p - \{E(X)\}^2$$

For the binomial distribution $B(n, p)$:

$$p_r = \binom{n}{r} p^r (1 - p)^{n-r}, \qquad \mu = np, \qquad \sigma^2 = np(1 - p)$$

For the Poisson distribution $Po(a)$:

$$p_r = e^{-a} \frac{a^r}{r!}, \qquad \mu = a, \qquad \sigma^2 = a$$

Continuous random variables

$$E(X) = \int xf(x)\, dx$$
$$\text{Var}(X) = \int x^2 f(x)\, dx - \{E(X)\}^2$$

Sampling and testing

Unbiased estimators:

$$\bar{x} = \frac{\Sigma x}{n}, \qquad s^2 = \frac{1}{n-1}\left(\Sigma x^2 - \frac{(\Sigma x)^2}{n}\right)$$

Central Limit Theorem:

$$\bar{X} \sim N\left(\mu, \frac{\sigma^2}{n}\right)$$

Approximate distribution of sample proportion:

$$N\left(p, \frac{p(1 - p)}{n}\right)$$

Sample exam papers

Paper 1

Q1 Andreas recorded the distances travelled from home to college by a random sample of 10 students from his college. The results, in kilometres, are given below.

 1.5   3.2   3.0   4.8   1.2   1.8   2.3   10.1   0.2   3.5

 (i) Find unbiased estimates of the population mean and variance. [3]

 (ii) State briefly what is meant by 'population' in this context. [1]

Q2 The heights of a certain kind of plant have mean μ cm and standard deviation 23 cm. In order to find a 95% confidence interval for μ, a botanist chooses a random sample of n trees. Find the least possible value of n so that the width of the confidence interval is less than 10. [4]

Q3 Samira wishes to choose a random sample of 600 people from a town of 21 300.

 (i) Explain fully how she can use random numbers to choose the sample. [3]

 It is known that 0.2% of the population have a certain medical condition.

 (ii) (a) Use a suitable approximating distribution to find the probability that more than 3 out of the 600 people in Samira's sample have this condition. [3]

 (b) Explain why your approximating distribution is appropriate in this context. [1]

Q4 The variable X is the number of people arriving at the Sales desk in a store every 5 minutes.

 (i) State two necessary conditions for X to have a Poisson distribution. [2]

 (ii) Given that X has the distribution Po(2.8), find the probability that exactly 4 people arrive at the Sales desk in a 12-minute period. [2]

 The store also has a Returns desk. The independent variable Y is the number of people arriving at the Returns desk every 10 minutes. It is given that Y has the distribution Po(1.3).

 (iii) Find the probability that a total of fewer than 3 people arrive at the two desks in a 3 min period. [3]

Q5 A machine makes drinks by mixing X cm^3 of juice with Y cm^3 of water, where X and Y are independent variables with $X \sim N(270, 7.5^2)$ and $Y \sim N(78, 4^2)$

 (i) The total volume of each drink should be greater than 330 cm^3. Find the percentage of drinks which have a volume less than this amount. [4]

(ii) Find the probability that, in a randomly chosen drink, the volume of juice is more than 3.5 times the volume of water. [5]

Q6 A random variable X has probability density function given by

$$f(x) = \begin{cases} k(3 - x) & 0 \leqslant x \leqslant 3, \\ 0 & \text{otherwise,} \end{cases}$$

where k is a constant.

(i) Show that $k = \frac{2}{9}$ [2]

(ii) Find the mean of X. [3]

(iii) Find a such that $P(X < a) = 0.19$. [3]

Q7 Weekly profits at a shop are normally distributed. In the past they had mean \$4150 and standard deviation \$370. New displays are introduced and over the next 15 weeks the mean weekly profit is \$4340.

(i) Stating a necessary assumption, test at the 2% significance level whether the mean weekly profit has increased. [6]

Later, a similar test at the 2% significance level was carried out using the profits from another 15 weeks.

(ii) Given that the mean weekly profit has increased, and is now \$4450, calculate the probability of a Type II error. [5]

Paper 2

Q1 The masses, in grams, of bags of sugar are represented by the random variable X with mean 504 g and variance 18 g^2. The total mass M, in grams, of a random sample of 5 bags is found. Find the mean and variance of M. [3]

Q2 X is a random variable having mean 12 and variance 5. The mean, $\overline{X}$, of a random sample of 50 values of X is taken.

(i) Find $P(\overline{X} > 12.5)$. [3]

(ii) State, with a reason, whether it was necessary to use the Central Limit theorem in part (i). [1]

Q3 The number of flaws per square metre in a certain type of cloth is modelled by a variable, X, having a Poisson distribution. In the past, the mean of X has been 6.4, but after a change to the manufacturing process, it is thought that the mean may have changed. A square metre of cloth is chosen at random and is found to contain exactly 2 flaws. Test at the 5% significance level whether the mean number of flaws per square metre has changed. [5]

Q4 Chris wants to estimate the proportion, p, of students at his college who own a computer. He chooses a random sample of 50 students from the college and finds that 42 own computers.

(i) Calculate a 90% confidence interval for p. [4]

(ii) The college principal claims that 95% of students at the college own a computer. Use your answer to part (i) to comment on this claim. [2]

Q5 At the last election, 92% of the population supported the president. A year later, a researcher wishes to test at the 10% significance level whether support for the president has decreased. She questions a random sample of 20 people and notes the number, X, who support the president.

   (i) State appropriate null and alternative hypotheses. [2]

   (ii) Assuming that the level of support for the president is unchanged, calculate $P(X \leqslant 17)$ and $P(X \leqslant 16)$ and hence find the probability of a Type I error. [4]

   (iii) If, in fact, only 75% of the population now support the president, calculate the probability of a Type II error. [2]

   (iv) Explain what is meant by a Type II error in this situation. [1]

Q6 The number of cars travelling north that arrive at a petrol station during a 5-minute period is represented by the variable X having the distribution Po(2.8).

   (i) Find $P(5 \leqslant X \leqslant 7)$. [2]

The number of cars travelling south that arrive at the petrol station during a 5-minute period is independent and has the distribution Po(0.4).

   (ii) A queue develops if a total of more than 4 cars arrive during a 5-minute period. Find the probability that this does not happen during any of the next twelve 5-minute periods. [3]

   (iii) Find the probability that a total of more than 340 cars arrive during an 8-hour period. [6]

Q7 The time, in hours, taken by cyclists to complete a course is modelled by the random variable X with probability distribution given by

$$f(x) = \begin{cases} \dfrac{k}{x^2} & 1 \leqslant x \leqslant 3, \\ 0 & \text{otherwise.} \end{cases}$$

   (i) Show that $k = \frac{3}{2}$ [2]

   (ii) Find the median of X. [3]

   (iii) Find Var(X). [6]

   (iv) Give a reason why the model may not be realistic. [1]

S2 Answers

The University of Cambridge Local Examinations Syndicate bears no responsibility for the example answers to questions taken from its past question papers which are contained in this publication.
When answers are not exact they have usually been given to 3 or 4 significant figures.

Chapter 1

Exercise 1a

1 (i) 0.217   (ii) 0.00625
   (iii) 0.879   (iv) 0.0338
2 (i) 0.0334 (ii) 0.967 (iii) 0.340 (iv) 0.256
3 (i) 0.434 (ii) 0.251 (iii) 0.497 (iv) 0.118
4 (i) 0.238   (ii) 0.567   (iii) 0.391
5 (i) 0.268   (ii) 0.269   (iii) 0.964
6 (i) 0.180 (ii) 0.0527 (iii) 0.135 (iv) 0.559
7 (i) 0.603   (ii) 0.178   (iii) 0.620
8 (i) Calls are independent of each other
   (ii) (a) 0.790   (b) 0.753   (c) 0.985
   (iii) 0.00318
9 (i) (a) 0.270   (b) 0.350   (c) 0.161
   (ii) 0.457
10 Breakdowns occur singly, randomly and independently.
   (i) 0.0959   (ii) 6
11 0.5
12 (i) Any two of the following: goals are scored randomly; goals are scored independently; goals are scored at a constant average rate.
   (ii) 0.155   (iii) 0.209

Exercise 1b

1 E-mails arrive independently
   (i) 0.0260   (ii) 0.697
2 (i) 0.224   (ii) 0.594   (iii) 0.472
3 (i) 0.983   (ii) 0.184   (iii) 0.199
4 (i) 0.0758   (ii) 0.223   (iii) 0.185
5 (i) 0.0821   (ii) 0.560   (iii) 0.0631
6 (i) (a) 0.380   (b) 0.616   (ii) 0.273
7 (i) 0.387   (ii) 0.929
   (iii) (a) 0.893   (b) 0.205   (c) 0.816
   (iv) 0.0288
8 (i) (a) 0.152   (b) 0.159   (c) 0.144
   (ii) 6
9 (i) 5.3   (ii) 0.102
10 (i) 5.29   (ii) 0.565   (iii) 0.0990
11 (i) 3   (ii) 2.45
12 (i) 0.590   (ii) 2.05
13 (i) 0.156   (ii) 0.0244
14 (i) 6.44   (ii) 4.61
15 3

Exercise 1c

1 (a) (i) 0.0476   (ii) 0.0498
   (b) (i) 0.225   (ii) 0.224
   (c) (i) 0.171   (ii) 0.168

2 (i) 0.0537   (ii) 0.0486
3 (i) (a) 0.184   (b) 0.0190
   (ii) 0.135   (iii) 0.0498
4 (i) 0.287   (ii) 0.251
5 (i) $\frac{1}{36}$ (ii) 0.713
6 0.305
7 0.109
8 (i) (a) 0.407   (b) 0.0629
   (ii) Events may not be independent as dialling wrong numbers may lead to more care being taken.
9 (i) (a) 0.0235   (b) 0.516   (ii) 0.954
10 (i) 0.185   (ii) 0.916
11 (i) 0.180   (ii) 0.538   (iii) 0.282
12 (i) B(300, 0.015)
   (ii) (a) 0.169   (b) 0.468   (iii) at least 200

Exercise 1d

1 (i) 64, 8   (ii) 0.0753   (iii) 0.9927
2 (i) 0.6201   (ii) 0.390   (iii) 0.5406
3 (i) 0.3998   (ii) 0.2004
   (iii) 0.3361   (iv) 0.0637
4 (i) 0.313   (ii) 0.5078
   (iii) 0.8335   (iv) 0.1101
5 (i) 0.383   (ii) 0.6561
6 (i) 0.2614   (ii) 0.2343   (iii) 0.0558
7 (i) 0.8901   (ii) 0.2377
8 (i) 0.1101   (ii) 0.7305
   (iii) 0.0019   (iv) 3.8 ≈ 4
9 (i) 0.4574   (ii) 0.8312   (iii) 0.0852
10 (i) 0.4594   (ii) 0.5363   (iii) 0.094
11 (i) 0.9815   (ii) 0.6247   (iii) 0.9244
12 (i) 0.0109   (ii) 0.8798
13 (i) 0.7832   (ii) 0.0731   (iii) 0.2074
14 (i) The number of faults per square metre has a constant mean; faults occur randomly; faults occur independently; faults occur singly
   (ii) (a) 0.202   (b) 0.692   (iii) 0.0109

Mixed Exercise 1

1 (i) 0.285   (ii) 0.310   (iii) 2.45
2 (i) 0.0902   (ii) 0.0613   (iii) 4.10 ≈ 4
3 (i) 0.144   (ii) 0.819   (iii) 2.88
4 (i) 0.224   (ii) 0.868
   (iii) No; visitors may not arrive singly or independently.
5 0.515

6 (i) 0.706 (ii) 0.987
7 (i) (a) 0.294 (b) 0.303 (ii) 1536
8 (i) 0.269 (ii) 6908
9 (i) 0.122 (ii) 0.532
   (iii) 0.0135 (iv) 0.229
10 (i) Any two of: customers buying the Daily Press arrive independently; customers arrive randomly; customers arrive singly; the mean rate at which the customers arrive is constant
   (ii) (a) 0.978 (b) 0.681 (iii) 0.022
11 (i) (a) 0.677 (b) 0.0527 (ii) 1498

Chapter 2

Exercise 2a

1 (i) 121, 144 (ii) 58, 36
   (iii) 6, 1 (iv) $-80, 64$
2 (i) $\frac{1}{4}$ (ii) 10
3 $a = 1.5, b = -10$
4 (i) 18 (ii) 5 (iii) 34
   (iv) 8 (v) 21 (vi) 7.68
5 34, 76.8
6 46, 12
7 (i) (a) $-3, 7$ (b) 25, 45
   (ii) (a) Poisson parameter cannot be negative; mean $\neq$ variance
   (b) mean $\neq$ variance
8 (i) 80, 12 (ii) 80, 6
9 100, 20
10 (i) 8, 0.949 (ii) 12, 0.6 (iii) 22, 1.407
11 300, 10

Exercise 2b

1 (i) 0.157 (ii) 0.0884 (iii) 0.798
2 (i) 0.857 (ii) 0.0248
3 (i) 0.0100 (ii) 0.126
4 (i) 0.329 (ii) 0.0231
5 (i) 6, 6 (ii) 0.285 (iii) 0.0923
6 (i) 0.110 (ii) 0.203
7 (i) 0.275 (ii) 6
8 (i) Any two of: people arrive independently, people arrive singly, people arrive randomly, the mean rate at which people arrive is constant
   (ii) 0.194 (iii) 0.00183

Exercise 2c

1 (i) 50, 4.61 (ii) N(50, 21.25)
   (iii) (a) 0.9850 (b) 0.1342
2 (i) 0.0239 (ii) 0.5585 (iii) 0.2849
3 (i) N (530, 13.25)
   (ii) (a) 0.9153 (b) 0.2775
4 (i) 243, 122.5 (ii) 0.0612 (iii) 261.2

5 (i) 0.0207
   (ii) (a) 0.0289 (b) 0.02 (c) 0.6252
6 (i) 0.0272 (ii) 0.002
7 (i) 10, 25 (ii) N(10, 625)
   (iii) 0.6554 (iv) 0.5793
8 (ii) 0.9324
9 (i) 0.1247 (ii) 0.6957 (iii) 0.7321
10 0.3124
11 (i) 0.1319 (ii) 0.0127
12 (i) 0.5 (ii) 0.8849 (iii) 0.2779
13 (i) 0.1728 (ii) 0.6127 (iii) 0.5
14 0.2575
15 (i) 0.1103 (ii) 0.753
16 (i) 9.6, 0.522
   (ii) (a) 1.8% (b) 22.2%

Exercise 2d

1 (i) N(60, 196) (ii) N(0, 1.361) (iii) N($-10, 49$)
2 (i) 0.8962 (ii) 0.9386
3 (i) 0.9441 (ii) 0.9441
4 0.0745
5 (i) 0.2399 (ii) 0.2074
6 0.6586
7 0.3242
8 (i) 0.8482, 0.9273
   (ii) Two cake sales; more chance of making money

Mixed Exercise 2

1 (i) 83, 15 (ii) 151, 13
   (iii) 83, 25.7 (iv) 205, 12.1
2 59.4, 7.09
3 (i) 0.0732 (ii) 0.377
4 (i) 0.962 (ii) 0.0915 (iii) 0.123
5 (i) 1.15 (ii) 0.216 (iii) 0.784 (iv) 0.776
6 (i) (a) 0.617 (b) 0.835 (ii) 0.0593
7 (i) 0.6298 (ii) 0.1056
8 (i) 450, 132 (ii) 0.6406
9 0.5863
10 (i) 0.8134 (ii) 0.5558
11 (i) 0.771 (ii) 0.8897
12 (i) 0.387 (ii) 10, 11.56 (iii) 0.6473

Chapter 3

Exercise 3a

1 (i) $\frac{3}{8}$ (ii) $\frac{7}{8}$ (iii) $\frac{13}{32}$
2 (i)

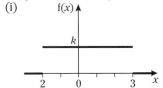

   (ii) 0.2 (iii) 0.74 (iv) 0.9
   (v) (a) 0.4 (b) 0.25

3 (i) 0.25 (ii) f(x)

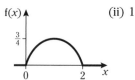

(iii) 0.375 (iv) 0.66

4 (i) $\frac{3}{56}$ (ii) $\frac{19}{56}, \frac{37}{56}$

5 $c = 1, k = 4$

6 (i) $\frac{1}{8}$ (ii) f(x)

(iii) $\frac{21}{64}$

7 (i) 0.64 (ii) 0.48

8 $k = \frac{1}{6}, a = 4$

9 (i) $\frac{3}{52}$ (ii) $\frac{19}{26}$

10 (i) $\frac{3}{2}$ (ii) $\frac{3}{4}$

11 $\frac{1}{4}$

12 (i) No, area $\neq 1$
(ii) No, f(x) < 0 for some values of x
(iii) Yes, area = 1, f(x) $\geqslant$ 0 for all x

13 $-\frac{3}{16}$

14 (i) 24 (ii) $\frac{8}{27}$

15 (ii) $\frac{3}{16}$

Exercise 3b

1 (i) 1.59 (ii) 1.26, 1.82 (iii) 0.557

2 (i) 1.76 (ii) 0.5

3 (i) 1.5 (ii) 1.2

4 (i) 6 (ii) f(x)

(iii) 2.16 (iv) 0.0330

5 0

6 (i) f(x) (ii) 0

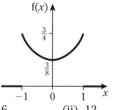

7 (i) 6 (ii) 12 (iii) 8

8 2.53

Exercise 3c

1 $4\frac{1}{3}$

2 2.27

3 $\frac{9}{16}$

4 $2\frac{23}{35}$

5 (i) f(x) (ii) 1

6 (i) 1.6 (ii) 0.4096
(iii) P(X < E(X)) < 0.5 and
P(X < median) = 0.5, so E(X) < median

7 (ii) 0.25

8 (i) $1\frac{2}{3}$ (ii) 1.74 (iii) 0.0981

9 (i) $k = \dfrac{1}{b-a}$ (ii) $a = 2, b = 14, k = \frac{1}{12}$
(iii) $\frac{1}{6}$

10 (ii) 1.5

11 (ii) $4\frac{2}{15}$

12 (ii) (a) $3p - 4q = 1$ (b) $p = 1, q = \frac{1}{2}$

Exercise 3d

1 (i) $2\frac{8}{15}$ (ii) 0.0822

2 (i) 1.607 (ii) 0.0742 (iii) 0.272

3 (i) 0.2 (ii) 0.5
(iii) $2\frac{1}{12}$ (iv) 1.44

4 (iii) 0.186

5 (ii) 0.2375

6 (ii) $\frac{50}{567}$

7 (ii) 0.4

8 (i) $\frac{1}{2}$ (ii) 0.869

9 (ii) 3 (iii) 3

Exercise 3e

1 (i) $2\frac{2}{3}$ (ii) $\frac{32}{3}\ln 2 - 7\frac{1}{9}$

2 (ii) $\ln\left(\dfrac{e+1}{2}\right)$ (iii) $\dfrac{1}{e-1}$

3 (ii) $\frac{1}{2}\pi$ (iii) (b) 0.75 (c) $\frac{1}{3}\pi$

4 (ii) $1\frac{7}{8}$ (iii) 1.906

5 (ii) $\tan^{-1}(0.5) = 0.464$ (iii) 0.399

6 (ii) $\dfrac{3}{2\ln 2} = 2.164...$
(iii) 0.727 (v) $2\sqrt{2}$

7 (i) 0.221 (ii) 0.607 (iii) 4 years

8 1.05

9 $\dfrac{1}{\ln(1.5)}$

10 (i) 4

11 (ii) $\dfrac{2 - \ln 2}{\ln 2}$

Mixed Exercise 3

1 (i) $\frac{1}{2}$ (ii) $\sqrt{2}$ (iii) $1\frac{1}{3}$

2 (ii) $\frac{2}{75}$ (iii) 0.931 (iv) 0.223

3 (i) $3\frac{1}{9}$ (ii) $\sqrt{10}$ (iii) $\frac{49}{144}$

4 (ii) 2.4 (iii) 1.2

5 (ii) 105 (iii) $5\frac{1}{7}$

6 (i) f(x)

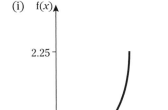

(iv) 0.822

7 (i) 10 min (ii) 5 min
(iii) 2.89 min (iv) 0.423

8 (ii) $25 = 30a + 9b$; $a = \frac{7}{12}$, $b = \frac{5}{6}$

9 (ii) 2.66 h (iii) 2.73 h (iv) 0.0243

10 (ii) 4.125 min (iii) $\frac{4}{27}$
(iv) less; $P(X > 5) < 0.25$

11 (ii) 70 ln(1.4) (iii) 0.528 (iv) greater

12 (i) $-\frac{3}{16}$ (ii) $\frac{19}{80}$

13 (i) (a) 0.451 (b) 0.449
(ii) 13.9 hours

14 (i) 1 (ii) $\frac{1}{2}\pi - 1$ (iii) $\frac{1}{6}\pi$

15 (ii) $\dfrac{7e^8 + 1}{2(e^8 - 1)}$

16 (i) 14.4 (iii) 0.281

Chapter 4

Exercise 4a

1 (i) In a census every member of the population is surveyed; in a sample members are selected from the population. A census is representative as it includes all the population. Using a sample is cheaper and less time-consuming.
(ii) Survey; quicker but care must be taken to avoid bias.

2 Assign each person a two-digit unique number (for example 01 to 50). Read off two digits at a time from the random numbers, ignoring repeats and numbers out of range, until 8 numbers have been selected. Choose the people corresponding to these numbers. Reading tables from left to right gives 02, 42, 38, 14, 12, 41, 10, 47 (other answers possible).

3 (i) Not representative as only union members have the chance to take part.
(ii) Take a random sample by acquiring a list of all employees and using the random number method to select the sample.

4 (i) Those without landlines will not be included; people in large households are less likely to be included than people in small households.

(ii) Obtain a list of all the adults in the town, number them, use a random number generator to select 25 numbers from the list and interview the corresponding adults.

5 (i) List the students and number them from 1 to 70.
(ii) 23, 45, 68, 3, 9, 43, 62, 42, 15, 33 (there are other correct lists)

Exercise 4b

1 (i) 0.0176 (ii) 0.2949

2 Assume sample is a random sample of all carrots sold by the wholesaler
(i) 0.9087 (ii) 0.9623 (iii) 0.8000

3 (i) 0.0828 (ii) 0.0832

4 (i) 0.3898 (ii) 6

5 (i) 0.2399 (ii) 0.0787
(iii) 0.0130 (iv) 109

6 251

7 3.49

8 17.3

Exercise 4c

1 (i) 0.1056 (ii) 0.3092

2 (i) 0.9848
(ii) Since n is large, by the Central Limit theorem, $\overline{X}$ is approximately normal.

3 (i) 0.7221 (ii) 0.1176

4 (i) 4.8, 2.88 (ii) $\overline{X} \sim N\left(4.8, \dfrac{2.88}{50}\right)$
(iii) (a) 0.4077 (b) 0.0704

5 0.0013

6 (i) 0 (ii) 0.4 (iii) 0, 0.008
(iv) $\overline{X} \sim N(0, 0.008)$ (v) 0.4873

7 (i) 0.9214
(ii) Yes; the distribution of the mass is unknown.

8 (i) $\overline{X} \sim N\left(5.8, \dfrac{5.8}{65}\right)$ (ii) 0.9434

Exercise 4d

1 236, 7.58

2 (i) 48.875, 6.98 (ii) 1.69, 7.36×10^{-6}
(iii) 22.8, 1.81 (iv) 33.6, 57.2

3 (i) 15, 43.1 (ii) 10, $3\frac{1}{9}$ (iii) 9.71, 621.1

4 0.5, 1.43

5 (i) 205.16, 9.22
(ii) greater than, since $s^2 = \dfrac{n}{n-1} \times$ sample variance and $\dfrac{n}{n-1} > 1$.

6 2.36

7 76, 145.455

8 (i) 748, 13.047 (ii) 0.0251

9 Assume random sample (or representative sample); 69, 16.203

Exercise 4e

1 (i) $(11.61, 12.79)$ (ii) 1.19
2 (i) $(139.49, 140.51)$ (ii) random sample
   (iii) 1.41 not in C. I.; claim not justified
3 (i) 12.45 (ii) $(10.75, 14.15)$ (iii) 3.4
4 (i) $(448.7, 467.3)$
   (ii) The probability that $(448.7, 467.3)$ includes μ is 0.99, on average, 99% of intervals calculated in this way will contain the true population mean.
   (iii) No; z-value less
5 (i) $(79.19, 84.81)$ (ii) $(78.90, 85.11)$
   (iii) No; n large, so $\overline{X}$ approximately normal
6 (i) $175.25, 103.5$ (ii) $(174.07, 176.43)$
   (iii) C. L. T. used to give approximate normal distribution for $\overline{X}$.
7 (i) $(51.25, 55.15)$
   (ii) 50 not inside C. I., claim not supported
8 (i) See page 96
   (ii) (a) $(68.03, 69.97)$
       (b) 1.94 (c) 0.077
9 $(747.3, 748.7)$
10 (i) $(1011, 1114)$ (ii) 36
11 (i) 5.06 (ii) 89%
12 47

Exercise 4f

1 (i) $(0.622, 0.738)$ (ii) 0.116
   (iii) $\sqrt{\dfrac{pq}{n}}$ has been estimated; distribution of P_s is approximately normal.
2 (i) $(0.293, 0.427)$ (ii) 0.175
3 $(0.0322, 0.143)$
4 (i) $(0.2637, 0.3897)$ (ii) 90
5 (i) 0.28 (ii) $(0.161, 0.399)$
   (iii) $\sqrt{\dfrac{pq}{n}}$ has been estimated; distribution of P_s is approximately normal.
6 (i) $(0.156, 0.344)$
   (ii) On average 3% of the intervals do not contain true p.
7 $(0.234, 0.346)$
8 (i) $(0.35, 0.49)$ (ii) 0.1368
   (iii) 1040
9 (i) $(0.223, 0.352)$
   (ii) Width is increased or C. I. is wider.
10 (i) Random sample; $(0.244, 0.283)$
   (ii) (a) 0.2635 (b) 90
11 $(0.50, 0.58)$
12 (i) large n (ii) $(0.043, 0.137)$
13 (i) 268 (ii) n smaller

Mixed Exercise 4

1 0.0195
2 0.9099

3 0.9351
4 (i) $9.5, 0.524$ (ii) 0.9717
5 (i) 0.324 (ii) 0.994
6 (i) $(30.37, 32.43)$ (ii) 2.06
7 (i) $(29.44, 32.56)$
   (ii) 30 within C. I.; evidence to accept claim at 2% level
8 (i) $375.3, 8.292$ (ii) $(0.133, 0.247)$
9 (i) $1050, 2304$ (ii) $(1030, 1070)$
   (iii) 246
10 (i) $227.1, 26.5$ (ii) 78
11 (i) $0.145, 600$ (ii) 97.4
12 (i) $(0.468, 0.612)$ (ii) 84.4
13 (i) 0.34
   (ii) $(0.30, 0.38)$, the probability that the interval includes p is 0.9, on average 90% of intervals would include true proportion
   (iii) 1520
14 (i) $9.49, 0.0129$ (ii) $(9.46, 9.52)$
   (iii) 71
   (iv) $\overline{X}$ becomes $\overline{X} + 0.05$ so answer to part (ii) should be amended; CI is shifted 0.05 units to right, $(9.51, 9.57)$; answer to part (iii) is unchanged as width uses s^2 which is not altered when all the readings are increased by 0.05

Chapter 5

Exercise 5a

1 $X \sim \text{B}(20, 0.3)$, $P(X \leqslant 3) = 0.1071 > 5\%$; not enough evidence of reduction
2 (i) $H_0: p = 0.65$, $H_1: p > 0.65$
   (ii) $X \geqslant 9$
   (iii) Accept Rachel's claim
   (iv) Yes; evidence to support Lily's claim
3 Do not reject H_0; not enough evidence to support manager's claim
4 (i) $H_0: p = \frac{1}{6}$, $H_1: p > \frac{1}{6}$ (ii) Not enough evidence of bias
5 Not enough evidence that success rate is less than 90%
6 (i) Do not reject H_0 (ii) Reject H_0
7 (i) Reject H_0 (ii) Accept H_0
   (iii) Reject H_0 (iv) Accept H_0
   (v) Reject H_0 (vi) Accept H_0
8 $P(X \geqslant 8) > 5\%$; not enough evidence that $p > 0.5$, so she could have obtained this score by guessing all the answers
9 $P(X \geqslant 8) < 6\%$; evidence to support Cycling Association's claim
10 Assume attempts are independent, evidence that Alex is losing his skills
11 $X \leqslant 1, X \geqslant 9$

12 (i) H_0: $p = 0.15$, H_1: $p < 0.15$, evidence of success
   (ii) Staff made an effort in the first week; take sample over a longer period

Exercise 5b

1 No evidence of decrease
2 $P(X \geq 4) = 0.054 > 5\%$, not enough evidence of increase
3 $P(X \leq 2) = 0.088 > 5\%$, not enough evidence of change
4 (i) $X \sim Po(3)$, $P(X \geq 5) > 5\%$, not enough evidence of increase
   (ii) $X \sim Po(9)$, $P(X \leq 3) < 5\%$, evidence of decrease
5 (i) $P(X \leq 4) = 0.0293 < 5\%$, reject H_0
   (ii) $P(X \leq 2) = 0.023 < 5\%$, reject H_0
   (iii) $P(X \leq 3) = 0.0818 > 4\%$, accept H_0
   (iv) $P(X \geq 4) = 0.0537 > 3\%$, accept H_0
6 Not enough evidence of decrease
7 3
8 H_0:$\lambda = 7.5$, H_1:$\lambda < 7.5$, $P(X \leq 3) = 0.0591 > 5\%$, not enough evidence of reduction
9 $X \geq 6$, i.e. at least 6 customers arrive

Exercise 5c

1 (i) $X \leq 1$
   (ii) You make a Type I error when you reject H_0 when H_0 is true
   (iii) 0.0385
2 (i) $X \leq 2$
   (ii) 0.803
3 (i) (a) Jessica would conclude that the coin is biased in favour of heads when it is fair.
      (b) 0.1938
   (ii) 0.276
4 (i) 0.0510   (ii) 0.697
5 (i) 0.062   (ii) 0.697
6 (i) $X \sim Po(7.2)$, $P(X \leq 3) = 0.0719 > 5\%$, not enough evidence of improvement in reliability
   (ii) Reject if $X \leq 2$,
      P(Type I error) $= P(X \leq 2) = 0.025$
7 (i) H_0:$\lambda = 4$, H_1:$\lambda < 4$, $P(X \leq 2) > 10\%$, not enough evidence of decrease
   (ii) (a) Saying that the rate of occurrence of flaws using the new procedure is unchanged when it has decreased.
      (b) $P(X \leq 1) < 10\%$, so reject if $X \leq 1$, accept if $X \geq 2$;
      P(Type II error) $= P(X \geq 2$ when $X \sim Po(2.4)) = 0.6916$

Mixed Exercise 5

1 (i) H_0: $p = 0.9$, H_1: $p < 0.9$
   (ii) Not enough evidence that proportion is less than 90%
   (iii) $X \leq 12$
2 (i) Defects occur randomly and independently
   (ii) $P(X \leq 6) = 0.0356 < 5\%$, evidence that mean number of defects has increased
3 (i) $X \leq 1$ (ii) 0.0477
4 4
5 (i) 0.37   (ii) 0.42
6 (i) (a) Rejecting H_0 when it is true
      (b) Accepting H_0 when it is false
   (ii) (a) 0.266   (b) 0.168
7 (i) Not a random selection
   (ii) one-tail, H_0: $p = 0.35$, H_1: $p > 0.35$, CR is $X \geq 6$, i.e. 6, 7, 8 survive, $x = 4$ is not in CR, so not enough evidence of improvement in survival rate
   (iii) A Type II error is made when you say that the survival rate is unchanged when there is an improvement
   (iv) 0.9502
8 (i) 0.0480
   (ii) 0.0480
   (iii) 0.601
9 (i) $P(X \geq 5) = 0.0959 < 10\%$; ploughing increases yield of metal
   (ii) Not enough evidence of an increase at 5% level
   (iii) 0.395
10 (i) 0.0202   (ii) 0.972   (iii) 0.0311

Chapter 6

Exercise 6a

1 (i) $z = 1.845$, reject H_0
   (ii) $z = 2.5$, reject H_0
   (iii) $z = -2.778$, reject H_0
   (iv) $z = -1.095$, accept H_0
2 $z = -0.943$, not enough evidence of decrease
3 $z = 3.069$, evidence that mean duration has increased
4 Not enough evidence that $\mu < 103.5$
5 $z = -1.66$, no change in mean
6 (i) H_0: $\mu = 6.0$, H_1: $\mu \neq 6.0$
   (ii) $5.778 < \bar{x} < 6.222$
7 $z = -1.878$, evidence that mean time is less, so training intensively has improved her times
8 (i) 33.95
   (ii) Evidence that mean speed is greater than 30 mph as sample mean is in the critical region

Exercise 6b

1   (i) 27.4
    (ii) $z = 1.987$, not enough evidence
2   (i) 1.70
    (ii) $z = -1.95$, evidence that mean time has decreased.
    (iii) By the CLT the distribution of the sample mean is approximately normal for large sample sizes.
3   $z = 1.909$, yes it does provide evidence
4   $z = 0.983$, accept that mean is zero
5   $z = 2.946$; yes, evidence that mean is higher
6   (i) 21.25
    (ii) $z = 0.99$, not enough evidence to support manufacturer's suspicion
    (iii) CLT used to obtain the distribution of the test statistic $\overline{X}$; necessary as distribution of X is unknown
7   (i) 10.46, 15.64   (ii) 1
    (iii) Do not reject H_0, not enough evidence that mean distance is greater than 10 km
    (iv) Since the sample size is large, the Central Limit theorem can be applied, so the distribution of mean distances, $\overline{X}$, is approximately normal

Exercise 6c

1   (i) Reject H_0 and conclude that the mean is not 52
    (ii) 0.04
2   (i) 0.0817   (ii) 0.665
3   (i) $\bar{x} < 122.5064$ (ii) 0.05
4   (ii) 0.24   (iii) $\mu > 389.7$ (iv) 0.0494
5   (i) 16.05 (2 d.p.)   (ii) (14.7, 17.3)
    (iii) 0.0684
6   (i) 0.0098   (ii) 0.7477
7   (i) 0.0606   (ii) 0.1118
8   (i) $H_0: \mu = 1.73$, $H_1: \mu > 1.73$
    (ii) $\overline{X} \sim N(1.73, 0.0008)$
    (iii) $\overline{X} > 1.777$
    (iv) Men who play basketball are not taller; assume players are a random sample of all basketball players
    (v) 0.14

Exercise 6d

1   (i) $z = 1.59$, accept H_0
    (ii) $z = 2.206$, reject H_0
    (iii) $z = -1.79$, accept H_0
    (iv) $z = 2.118$, accept H_0
    (v) $z = -2.937$, reject H_0
2   $z = 1.705$, evidence that more than 65% own a cell phone

3   $z = -2.40$, manufacturer's claim is not supported as there is evidence that proportion is less
4   $z = 1.637$, evidence that support has increased
5   $z = 1.476$, not enough evidence that college population is not equally divided
6   $z = -1.990$, not enough evidence that proportion is less
7   $z = 1.5$, accept that proportion is 0.1
8   (i) (a) 0.0297   (b) 0.0934
    (ii) $z = -1.792$, evidence that Everyday seeds have a germination rate less than 75% (but only just, so do further tests)
9   $z = 2.43$, evidence that more than a third belong to a youth organisation
10 (i) Assume replies were representative of all teachers; $z = 1.220$, do not reject H_0, not enough evidence that proportion in favour is more than 0.7
    (ii) (0.681, 0.808)
11 $z = -1.267$, no; not enough evidence to reject that $p = 0.75$
12 (i) Not a random sample; not representative as it excludes people at work, school, etc at this time; take random samples at random times during the day over several days.
    (ii) $z = -2.03$, reject H_0, so data do provide evidence that percentage is less than 80%
13 (i) 68
    (ii) 0.0948
14 (i) (a) $H_0: p = 0.13$, $H_1: p < 0.13$
        (b) 2%
    (ii) 0.161

Exercise 6e

1   (i) $z = 1.355$, accept H_0
    (ii) $z = -1.447$, reject H_0
    (iii) $z = -2.5$, reject H_0
    (iv) $z = 0.9095$, accept H_0
    (v) $z = -2.0156$, reject H_0
2   $z = 2.055$, not enough evidence of increase
3   $z = -2.45$, evidence of decrease
4   $z = 1.296$, not enough evidence of increase
5   $z = -2.628$, evidence that mean is not 4.2
6   (i) 0.106
    (ii) 0.6168
7   $z = 2.517$, evidence of increase

Mixed Exercise 6

1   $z = -1.47$, not enough evidence that mean has changed
2   (i) $H_0: \mu = 46$, $H_1: \mu \neq 46$
    (ii) Not enough evidence at 5% level that mean time has changed

3 $z = 2.553$, evidence to support manufacturer's claim

4 (i) 0.985
  (ii) $z = -0.740$, accept H_0, no significant change

5 (i) $z = 2.77$, evidence that mean amount of the chemical in the bloodstream of patients taking drug B is different from that of patients taking drug A.
  (ii) No: since n is large, by Central Limit theorem $\overline{X}$ is approximately normal

6 $z = 1.567$, not enough evidence to say that the quoted figure is an underestimate

7 (i) 66.25, 133.40
  (ii) H_0: $\mu = 62.50$, H_1: $\mu > 62.50$, $z = 1.465$, not enough evidence of increase

8 (i) $z = -3.03$, evidence that $p < 0.4$
  (ii) (0.379, 0.458)
  (iii) 75

9 (i) 52.74
  (ii) 0.103

10 (i) 6.525, 2.87
  (ii) $z = -2.52$, evidence of reduction in mean number of cars caught speeding
  (iii) Saying that there is a reduction in the mean number of cars caught speeding when there is not a reduction

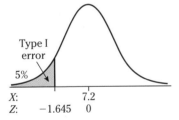

11 (i) H_0: $\mu = 3$, H_1: $\mu > 3$, $z = 1.43$, not enough evidence to support the claim
  (ii) Saying there is no extra weight loss when there is

12 $z = 1.327$, evidence of increase

13 (i) 14.8, 9.80
  (ii) H_0: $\mu = 15.2$, H_1: $\mu < 15.2$; 0.1; saying the photographer has fewer discards when she doesn't
  (iii) $z = -0.915$, not enough evidence to support photographer's claim

14 (i) $\overline{x} < \mu_0 - 1.96\dfrac{\sigma}{\sqrt{n}}$ or $\overline{x} > \mu_0 + 1.96\dfrac{\sigma}{\sqrt{n}}$
  (ii) $\overline{x} > \mu_0 + 2.326\dfrac{\sigma}{\sqrt{n}}$

Sample Paper 1

1 (i) 3.16, 7.66
  (ii) Students at Andreas' college.

2 82

3 (i) Number people $1-21\,300$. Generate 600 valid 5-digit random numbers, ignoring repeats and numbers over $21\,300$. Select the corresponding people.
  (ii) (a) 0.0338
    (b) Binomial with $n = 600 > 50$ and $np = 1.2 < 5$; hence use Po(1.2).

4 (i) People arrive randomly, singly and at constant mean rate (any two).
  (ii) 0.103
  (iii) 0.658

5 (i) 1.71%   (ii) 0.425

6 (ii) 1   (iii) 0.3

7 (i) $z = 1.989$; cf 2.054; there is no evidence at the 2% level that mean weekly profit has increased.
  (ii) 0.139

Sample Paper 2

1 2520, 90

2 (i) 0.0570
  (ii) Yes. Distribution of X unknown, so need to use the fact that distribution of $\overline{X}$ is approximately normal for large sample size.

3 $p = 0.0463 > 0.025$. There is no evidence at the 5% level that the mean number of flaws per square metre has changed.

4 (i) (0.755, 0.925)
  (ii) Claim not justified as 95% lies outside interval.

5 (i) H_0: Support in population is 92%; H_1: Support in population is less than 92%
  (ii) 0.0706, 0.212, 0.0706
  (iii) 0.225
  (iv) Concluding that support has not decreased when it actually has

6 (i) 0.144
  (ii) 0.0512
  (iii) 0.0287

7 (ii) $\dfrac{3}{2}$ (iii) 0.284
  (iv) For example, it does not allow for times below 1 h or above 3 h.

Index

Acknowledgements

Previous exam questions from Cambridge International AS and A Level Mathematics 9709 reproduced by permission of the University of Cambridge Local Examinations Syndicate:

Page 5, Example 1.5: paper 7 question 6 November 2005; Page 8, Example 1.7: paper 7 question 6 June 2004; Page 13, Example 1.10: paper 7 question 4 November 2003; Page 17, Example 1.13: paper 7 question 2 November 2002; Page 21, Example 1.16: paper 7 question 5 June 2007; Page 24, Mixed Exercise 1, Q3: paper 7 question 6 June 2008, Mixed Exercise 1, Q5: paper 71 question 1 November 2009, Mixed Exercise 1, Q8: paper 7 question 5 November 2004; Page 25, Mixed Exercise 1, Q9: paper 7 question 6 June 2006; Page 27, Example 2.3: paper 7 question 1 June 2005; Page 29, Example 2.6: paper 7 question 3 June 2004; Page 34, Example 2.9: paper 7 question 1 November 2004; Page 35, Example 2.11: paper 71 question 6 June 2010; Page 42, Example 2.18: paper 7 question 4 June 2006; Page 45, Example 2.21: paper 7 question 3 November 2008; Page 49, Example 2.24: paper 71 question 4 June 2010; Page 50, Example 2.25: paper 7 question 4 November 2007; Page 53/4, Example 2.27: paper 7 question 5 November 2006; Page 57, Mixed Exercise 2, Q2: paper 7 question 2 June 2006, Mixed Exercise 2, Q4: paper 7 question 6 June 2003, Mixed Exercise 2, Q5: paper 7 question 6 November 2008, Mixed Exercise 2, Q6: paper 7 question 6 June 2005; Page 58, Mixed Exercise 2, Q9: paper 7 question 3 November 2003, Mixed Exercise 2, Q10: paper 7 question 2 November 2002, Mixed Exercise 2, Q11: paper 71 question 7 November 2009, Mixed Exercise 2, Q12: paper 7 question 7 November 2005; Page 74/5, Example 3.12: paper 7 question 7 November 2006; Page 81, Example 3.15: paper 7 question 6 November 2004; Page 84, Example 3.17: paper 71 question 5 June 2010; Page 85, Example 3.18: paper 71 question 5 November 2009; Page 87, Example 3.19: paper 7 question 7 June 2008; Page 88, Example 3.20: paper 7 question 7 November 2003; Page 92, Mixed Exercise 3, Q2: paper 7 question 5 June 2006, Mixed Exercise 3, Q3: paper 73 question 5 November 2010, Mixed Exercise 3, Q5: paper 7 question 7 June 2005 , Mixed Exercise 3, Q6: paper 7 question 7 June 2007, Mixed Exercise 3, Q9: paper 7 question 7 November 2008; Page 93, Mixed Exercise 3, Q10: paper 71 question 5 June 2009, Mixed Exercise 3, Q11: paper 7 question 6 November 2002; Page 97, Example 4.1: paper 73 question 2 November 2010; Page 100, Example 4.2: paper 7 question 2 June 2008; Page 105, Example 4.5: paper 7 question 1 November 2004; Page 110, Example 4.9: paper 7 question 3 June 2008; Page 116, Example 4.12: paper 73 question 4 November 2010; Page 118, Example 4.14: paper 7 question 3 November 2005; Page 125, Example 4.17: paper 7 question 3 November 2006; Page 126, Example 4.18: paper 7 question 3 June 2005; Page 127, Example 4.19: paper 7 question 1 November 2007; Page 131, Mixed Exercise 4, Q4: paper 7 question 2 June 2004, Mixed Exercise 4, Q5: paper 71 question 6 June 2009, Mixed Exercise 4, Q7: paper 7 question 3 June 2003, Mixed Exercise 4, Q8: paper 7 question 4 June 2004; Page 132, Mixed Exercise 4, Q9: paper 7 question 6 June 2007, Mixed Exercise 4, Q10: paper 71 question 2 June 2009, Mixed Exercise 4, Q11: paper 71 question 2 June 2010; Page 138, Example 5.2: paper 7 question 1 November 2007; Page 149, Example 5.7: paper 7 question 2 November 2005; Page 151, Example 5.9: paper 71 question 4 November 2009; Page 152, Example 5.12: paper 71 question 4 June 2009; Page 154, Example 5.13: paper 71 question 7 June 2010; Page 155, Example 5.14: paper 73 question 7 November 2010; Page 159, Mixed Exercise 5, Q3: paper 7 question 4 November 2002 , Mixed Exercise 5, Q6: paper 7 question 6 November 2003, Mixed Exercise 5, Q7: paper 7 question 7 November 2004; Page 160, Mixed Exercise 5, Q8: paper 7 question 4 June 2007, Mixed Exercise 5, Q9: paper 7 question 6 November , Mixed Exercise 5, Q10: paper 7 question 5 November 2008; Page 169, Example 6.2: paper 71 question 1 June 2009; Page 171, Example 6.4: paper 7 question 2 November 2008; Page 177, Example 6.8: paper 7 question 3 November 2002; Page 180, Example 6.9: paper 7 question 4 June 2005; Page 181, Example 6.10: paper 7 question 5 June 2003; Page 191, Example 6.15: paper 7 question 3 November 2005; Page 196, Mixed Exercise 6, Q1: paper 7 question 3 June 2007, Mixed Exercise 6, Q2: paper 7 question 1 November 2006, Mixed Exercise 6, Q4: paper 7 question 5 November 2003, Mixed Exercise 6, Q5: paper 73 question 6 November 2010; Page 197, Mixed Exercise 6, Q9: paper 7 question 5 June 2004, Mixed Exercise 6, Q10: paper 7 question 7 June 2006, Mixed Exercise 6, Q11: paper 7 question 4 June 2008; Page 198, Mixed Exercise 6, Q13: paper 71 question 6 November 2009.